The Interaction
of Science
and Technology

A Symposium held at the
University of Illinois, Urbana
October 17 and 18, 1967

Speakers

EMANUEL R. PIORE

JACOB E. GOLDMAN

CHALMERS W. SHERWIN

WILLIAM J. PRICE

MORRIS TANENBAUM

DANIEL ALPERT

GEORGE E. PAKE

WILLIAM K. LINVILL

THE INTERACTION
OF SCIENCE
AND TECHNOLOGY

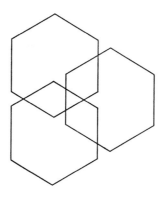

W. Dale Compton, Editor

UNIVERSITY OF ILLINOIS PRESS
Urbana   Chicago   London

Preface

American science, faced with a curtailment in its funding and with a decreasing appeal to new students, has begrudgingly come to realize that the "research honeymoon" is ending. The scientific community is now expected to give substantive answers to such queries as what fraction of the GNP should go toward the support of science and technology, is the support of science already excessive, what areas in science and engineering should be singled out for increased support, by what justification do mission-oriented agencies support basic research, etc.

It is not uncommon for many of us, in justifying the investment in basic scientific research, to point out that technology feeds on the results of research and thus that the state of our technology justifies our present rate of investment in science. Since we have obviously been successful, it is argued that a further increase in the rate of expenditure for basic science would clearly yield increased benefits.

Such arguments appear to be somewhat less than convincing to the majority of the taxpayers and to the Congress. Although few are willing to seriously deny the technological eminence of this nation, many seem to doubt that either an *increasing rate* or maintenance of the previous rate of in-

crease of six to ten per cent per year of investment in fundamental science will *guarantee* an equally increased rate of
technological development, particularly when the nation is
faced with finding the means to solve the massive problems
of the city, of poverty, and of equal opportunity for minority
groups.

Clearly, however, there is some connection between
technological achievement and the products of basic research.
In its simplest form the question might be stated as, what is
this connection and how can it be strengthened? This
thought was the central theme of the Symposium on the
Interaction of Science and Technology. The Symposium
brought together a small group of eminent physical scientists
from the industrial, governmental, and university communities to discuss the circumstances in which they had observed a successful interaction between the scientist interested
in basic research and the scientist or engineer interested in
transferring the results of this research to practice. The papers presented in this volume constituted the formal presentation at the Symposium. Chairmen of the sessions and
leaders of the discussions were Dr. Herbert Carter, Vice-
Chancellor of Academic Affairs, University of Illinois, Urbana; Dr. Donald Hornig, Scientific Adviser to the President;
and Dr. Harvey Brooks, Dean of Applied Sciences at Harvard
University. The occasion of the Symposium was the dedica-
tion of the University of Illinois' Coordinated Science Laboratory Building with its Space Science Wing. Mr. James E.
Webb, Administrator of the National Aeronautics and Space
Administration, presented the address at the dedicatory
luncheon and participated in the Symposium.

Sponsorship of the Symposium was provided by the
Office of Naval Research. The assistance of Mr. Donald
Pollack, of that office, is gratefully acknowledged.

W. Dale Compton
Urbana, Illinois

Contents

1 EMANUEL R. PIORE

Science
and Technology
in Industry

The present role of science and technology in industry has historical roots. To understand the contemporary atmosphere and problems, I want to spend a few minutes reciting or interpreting this history. This will be a narrow interpretation and I will confine myself to electronics and to this country.

Prior to the war—1940—there was a handful of firms that had what we would recognize as research laboratories, distinct units functioning as research organizations: Bell Telephone Laboratories, General Electric, Westinghouse. During the war many physicists left their laboratories and tackled engineering problems with great success and acclaim. This created a point of view which became widely accepted, the point of view that great engineering accomplishments come from research. Parenthetically, let me observe that the war effort did demonstrate that those who were well grounded in the basic sciences could make important and, at times, unique contributions to engineering.

The generation that was involved in the war is now the

Emanuel R. Piore: Vice-President, Chief Scientist, and member of the Board of Directors, International Business Machines Corp., New York.

source of our elder statesmen; some were young enough then so that they are now in their prime while others are retired or about to retire. This group of researchers returned to their academic activity with an exposure to engineering, with experience in how to do engineering, in how to bend with the wishes and needs of the users, and with the beginning of an understanding of the interaction of society and technology, at times calling it society and science.

After the war, in the late 1940's and early 1950's, many other large industrial firms started new organizational entities dedicated to research. General Motors, Ford, and U.S. Steel are examples. Obviously, there were also many electronics firms. Underneath this large expansion in industrial research was the notion, very seldom articulated explicitly, that new products come about through an orderly procedure starting with science, moving through applied research, and finally appearing as some manufactured article, system, or new structure that is viable in the marketplace and has a beneficial impact on our society.

This point of view also prevailed in academic circles and took two forms that I want to identify. First, there was a profound shift in engineering school curricula, a shift to include more fundamentals, more science, and to eliminate courses such as drafting. Secondly, there was the evolution of a national science policy, which stated in its simplest form was: support scientific research and the ills of our society will be solved; also, the most efficient place to do research is in a university environment.

Now, after 20 years of continued expansion of research in universities, government, and industry, serious questions are being raised. Research has not produced all the anticipated benefits, at least not visible benefits for society as a whole. There has also been a restlessness among the engineers who wanted recognition. This led to the National

Academy of Engineering. Engineers were restless also with the growth in the sciences and wanted systems approaches, case studies, etc. In addition to this reaction in the engineering community, there have been more disturbing reactions in industry and in government, where we've seen new statements of science policy. Even in the scientific community, voices have echoed this theme that science, or research, has failed society. The Department of Defense, an important source of funds for universities and industry, takes the Hindsight Study seriously and is slowly but surely eliminating support in many important areas of physics and chemistry. The White House issues statements that we must now apply the results of our research. There is a body of scientific opinion to the effect that our selection of research programs should be based on predictable benefits to society.

In going through this history, I have tried to be neutral and not expose my prejudices. I do have certain firm convictions that I will now articulate, but in order to get understanding and support for these convictions, it is necessary to understand the attitudes of one's peers and of those who supply the funds.

Large industrial companies have a spectrum of technical activities that extend from basic research to production. The sociologists of science classify these activities in a variety of ways by the use of phrases such as applied research, advanced development, development, production engineering, manufacturing engineering, and manufacturing. Points on this spectrum are best characterized by the freedom a technical person has in selecting his problem. The basic research person supposedly has the greatest freedom to select his problem. The technical person on the manufacturing floor supposedly has the least freedom. Very broadly, they are both in the same technical environment. They differ in their motivation, in their audience, in their rewards, recognition, and personal satisfaction. The motivation for the person's involvement is the rewards he receives from society or the audience to which he addresses himself.

I approach this spectrum of technical work on the basis that all the people involved work on real problems which have intellectual content, and that a high degree of training is required for the solution of those problems. I am not trying to create a hierarchy of snobbishness. A research-oriented person generally looks to the outside world for his audience and his recognition, not necessarily within the social structure in which he works. This is true in universities, government, and industry. A research person gets greater satisfaction from recognition from his peers with whom he does not live than from his peers who reside in his immediate vicinity. In spite of this, industry in our contemporary world does want research people in its laboratories and it makes certain demands on them—demands, most simply stated, for responsiveness and sensitivity to the technical environment of that industry; their choice of problems should be motivated by the man-made world of the industrial complex in which they work. One hopes that in observing the fabrication process they will find important problems, whether in development or in product engineering, and, simultaneous with their coupling to the outside world, they will do research that may have an impact on future products, processes, or systems within a given industrial complex.

Normally, the sociologists of science and technology concentrate on the last item and do not quite realize the importance of the total research operation and what are considered its less glamorous functions. Many manufacturing procedures are based on the state of the art. This is especially true in our contemporary world in the fabrication of transistors or more complex structures known in the trade as monolithics. There is a large element of alchemy, of trial and error, of "black magic." There is a lack of fundamental understanding of the processes involved. From time to time, these processes go bad and one is at a loss to find the reason. The normal procedure is to stop the line, clean everything thoroughly, paint everything thoroughly, and then start all over again. Very often everything goes well after the cleaning process. This obviously produces an element of uncertainty and discomfort, a certain nervousness. Having been stimulated by the existence of this feeling of panic, one looks to research to try to give some fundamental understanding so that one can gain control, so that one can better understand the man-made world.

Quite apart from these catastrophic considerations, one can observe historically that, as one develops an understanding of a process and anchors it to first scientific principles, one acquires an opportunity to reduce costs. I will continue to stress this function of research in industry because it is the least understood by those who write about it. Normally we read about the orderly process of something coming out of basic research, going through applied research, and finally, through the other categories I have identified, into manufacturing. Our world is not that orderly.

We also design products and systems based on past experience, past products, intuition, and general knowledgeability. One can cite many historical examples. Coming back to the contemporary world to an area in which I have direct contact, one finds that computers—their structure, the architecture of the total system—evolve from experience, esthetic appeal, intuition, and some understanding of the problems that the computer will solve. This is true not only of the hardware configuration of the computer but also of the software. We are very far from starting with first principles of science, going through careful analytic procedures, and coming up with a design for the hardware and software of a computer. Progress is being made and some general theorems are being evolved, but we remain very far from the goal.

Thus, it is hoped that a person dedicated to research will be stimulated by these large gaps in fundamental understanding and will dedicate himself to trying to obtain better understanding, to see whether he can find first principles and determine whether a systematic intellectual base can be built. The mysteries of our world reside in some of the very structures we have built. Not all mysteries reside in the stars and in the natural systems.

There are some standard historical examples of this and I might as well throw them out. We had a great deal of communication going on before we began to put a structure under it with information theory. We had many steam engines serving and revolutionizing society in railroads and steamships before we had adequate structures known as thermodynamics and statistical mechanics.

What one wants from a research person is *a reaction to his environment, taste in the problems he chooses, and a great impatience with trivia.* Thus one provides the stimulus and gives the man the freedom to respond to it. Unfortunately, there are very few people who have the gifts to do research. And our universities have drilled into their graduate students that if they don't do research, especially basic research, they somehow have failed in their purpose on this earth. And they do not learn the questions: is this a good problem? is it a serious problem? is it a solvable problem? So one of the first judgments industry must make as a young man joins it is: has he the taste and talent to do research? can he be given freedom? can he be productive—productive to society while getting satisfaction out of his work? This sort of judgment is very difficult and produces much irritation and aggravation.

The function of research in industry is to look at the

totality of technical activities in a given firm and, from that totality, select areas that are amiable to solution and illumination. The validity of this characterization of research in industry can be observed in the mobility of people. Some time is spent in research, then there is a drift into development, and at times into the more commercial function. In contrast, a similar drift in academia presents a different pattern—department, dean, laboratory director, etc.

Research in the universities has a much more restricted set of stimuli as the scientist tries to look at the man-made world. This is why certain areas of research are dominated by industry.

It is very difficult for me to differentiate between basic research and applied research. I guess all the work currently going on in holography can be classified as applied research. Early laser work can possibly be classified as basic research, and much of the present laser work can be classified as applied research. I do not quite understand these classifications. I guess applied research is associated with making some device work or applying some principle without necessarily committing oneself to a specific system or specific product or black box. Possibly the difference is whether one publishes in the *Physical Review* or in the *Journal of Applied Physics*. The research person feels rewarded if he publishes on the outside and is acclaimed by his peers. The applied research person, I guess, hopes that some of his ideas actually will end up in some useful gadget, some useful application that will hit the marketplace, whether a commercial marketplace, a military marketplace, or a social marketplace. One finds that possibly one-third of the applied research finds immediate utility, for other considerations come into play before it can hit the marketplace. These considerations include the existence of a marketplace and the cost of the gadget, black box, or system. This interaction of technology and society is another topic.

I have commented on the following matters: the broad interests of those academic men who participated in the war, the questioning of science as the fountainhead of the solutions for problems facing society, the lack of agreement on what training to provide an engineer, and the role of research in industry.

These apparently diverse observations are related, and now I should like to draw some conclusions which I will state as problems.

How can one create the state of mind that a very productive, rewarding, and creative life can be led even though one is not a theoretical physicist? How does one erode the highway of intellectual snobbishness? How does one present the functions of research in the world? It is all-pervading and prevailing.

How can we make our young researcher aware of those problems our society faces to which he can make a contribution through his training and acquired skills while receiving satisfaction and the recognition of his peers? There are many volunteers to modify our foreign policy and social structure. A technically trained person by his training has no unique wisdom in these areas, but he can contribute in an area that requires his skills.

JACOB E. GOLDMAN

Science for Economic Growth and Social Change

First came the ivory tower, then the corridors of business and commerce, then government and the apparatus of war and now—with sudden impact—society at large. This has been the progression of acceptability and utilization of scientific research.

Until relatively recently, scientific research (as distinct from technological or engineering research) was the exclusive province of the academic campus. Its utility was considered at best vague and the thread that linked its practice to the evolution of new technologies so winding—both spatially and temporally—that no one felt any compulsion to intrude upon the privacy and sanctity of the researcher. Research was (and is) synonymous with acquisition of new knowledge. Where but at a university should such aspirations find a home?

Then came the period between the two great wars—the period of great industrial expansion in the United States. Here and there it was discovered that science can be profitable in leading to and pointing up new materials, methods,

Jacob E. Goldman: Director, Scientific Laboratory, Ford Motor Company.

and devices and thus bridge the gap—both in time and in space—between scientific discovery and technological invention. Research suddenly became a corporate asset—an intangible one, but of sufficient reality to be recognized by the more progressive and aggressive corporate managements.

During World War II and its aftermath, the government joined the parade. It realized that science can be not only an industrial asset generating economic opportunities but an essential national asset in maintaining the security of the nation. Military weaponry and national prestige—particularly in space exploration—grew to depend more and more upon the nation's scientific capability. Government suddenly realized its responsibility not only to exploit the fruits of science in these areas of national interest but to assure the maintenance of the capability—survival of the species. Government thus became the principal supporter of science and of the scientists who reside in their native habitat of the campus, in the more recently acquired environment of industry and in the in-house apparatus created by the government for this specific purpose.

The impact of the wartime and postwar triumphs of science has more recently brought a new dimension into this saga of the care and feeding of science and the scientist. The bomb, radar, the conquest of space and the emergence of the computer have made it clear that science touches upon the everyday life of every man. Aided and abetted by communications media, concern for science in all its manifestations has spread to the public at large. Science is written about daily in the public press; there are even several comic strips devoted to the serious expostulation of matters scientific and some not-so-serious ones that will tell you that whoever masters magnetism will rule the world; it is talked about daily in the legislative bodies at national, state, and local levels; it has even conferred a sort of prestigious status on its practitioners. The scientist in the public eye is no longer the mad iconoclast but a normal human being.

But this spreading of the informational and financial support base has brought with it a whole new series of problems and challenges. It is to some of these that I choose to devote the balance of my remarks today.

A question being asked again and again these days is: Why can't the science that created nuclear energy, that put a man in space, that fashioned incredibly fast and powerful computers—why can't this scientific apparatus be put to work to solve our major social problems? If, indeed, the techniques of science are so powerful, so all embracing—goes the query—why can't they be put to work improving the lot of man—putting an end to the major social difficulties confronting us? All we have to do is to institutionalize the process, a la the Manhattan project, and all the ills of the world will disappear through the magic of science.

What are some of these critical social problems that need concerted attack? Any listing of these will usually include health (cancer, heart, and stroke), crime, waste disposal, urban blight (housing and construction), pollution, and transportation.

Since two of these problem areas have been of more than tangential concern to the scientific organization with which I am associated, and some of the contributions which we have made toward understanding these problems—pollution and transportation—are hopefully not entirely trivial (one of these that I will talk about has been in collaboration with the very laboratory whose facilities we are dedicating today), I will offer them as meager but possible credentials for discussing some ideas concerning the general problem. The examples I cite are there because I believe they are relevant and permit some extrapolation to other areas of concern; if they inject an air of rank commercialism or inadvertent corporate boastfulness, I apologize in advance.

As the crusading forces for social change gather momentum and pounce with missionary zeal on all sorts of panaceas, it has become evident that solutions are sought for a problem—or, more properly, a series of problems—that are ill defined and little understood. Unfortunately, there is no panacea. As science graduates from a stage in which its progeny are strictly technological to one in which its effects have direct and major social consequences, the method and institutionalization become somewhat more complex. Not only does much of the science that will be needed to help solve these problems not yet exist but it has not yet been put to work to properly define the problems. Moreover, it is not clear that, as solutions emerge, the problems will remain the same.

Let me cite as an example the case of transportation. This is often cited as one of those instances where modern society has boxed itself in. In urban areas we have congestion; in suburban areas, waste or poor economics or bad planning; and in rural areas the problem of providing rights of way, modal interfaces, etc. The critics ask why science and technology can't straighten this situation out. It is clear, however, that science in endeavoring to provide solutions enters a sphere where an uncertainty relation of sorts prevails: the process of providing a solution perturbs the very problem for which a solution is being sought. A transportation system modifies the demography of the macroregion in which it is placed, just as a new drug useful in combating a health menace may alter quite perceptibly the ecology of the species in unpredictable ways.

I pose, therefore, the following questions. Can science afford to fall back on its classical detachment from its application; can it continue to remain orthogonal to the social values, relying on distillation by others to produce interaction with the outside world? Or is it necessary, in order for science to be useful in this new dimension, to somehow create new mechanisms for interaction or new institutions for discharging its social mandate?

The answers, I believe, lie not in changing the pattern of scientific research per se but in the concept that has suddenly become "in" but has been practiced for many years in industrial laboratories—coupling.

Industry has learned to compress the time scale between discovery and invention—between new science and profitable technology—by bringing together those whose interests lie within the bounds of pure science and those whose personal and professional goals are more oriented toward the broader technical goals of the company. Basic science is necessarily discipline oriented and must remain so; applied science is by nature interdisciplinary. If science is to be useful in the solution of specific problems, be they technical, economic or social, there must be a milieu in which science can thrive in all its disciplinary grandeur—much as it does at a university or at an industrial institute where there resides a proper appreciation and understanding of the ground rules under which it thrives. At the same time, there must be provided an environment of technological awareness conducive to its interaction with the outside world. It must be insulated from the day to day involvements of the marketplace but not isolated from the needs of the marketplace. It requires a milieu of technological awareness.

There can be no doubt that American industry has learned to take full advantage of science in precisely this way. It may be that this factor has played a significant—perhaps even dominant—role in creating the celebrated "technology gap." Industrial scientific research as we know it in America has virtually no counterpart in Europe or elsewhere.

What of science-based social change? Are there analogous principles that can be put to use? Two examples from my own immediate experience come immediately to mind: Let me talk first about pollution. In a recent report the President's Science Advisory Committee identified air and water pollution as one of society's most critical problem areas—an inevitable consequence of the burgeoning industrialization and urbanization of this society.[1] Upon the automobile was placed a significant portion of the blame for polluting the atmosphere. To the citizens of London and Los Angeles, this was no startling revelation. But for them and for the rest of the country it helped quantify an existing problem of increasing moment.

What solutions can be envisioned either below or beyond the horizon? Do we replace the automobile with other transport modes in downtown areas? If we do, we might solve one problem only to create a host of others far more pressing to our economy and our society. Do we replace the internal combustion engine with a nonpolluting prime mover? Once again, unless science can provide a fully competitive energy source, society might not accept the major compromises demanded by such a substitution. Do we clean up the gasoline engine? This might be an ideal solution, but the science and technology to tell us how to do this does not yet exist.

It is not pure coincidence that two of the most promising avenues toward ultimate solution come from the very type of scientific environment I mentioned a moment ago. As far as I know, the only battery that even comes close to providing the energy and power requirements for a power plant competitive with the internal combustion engine is the one invented in our laboratory—not by technologists bent on solving the pollution problem—but by a physical chemist concerned with the atomistic nature of catalytic surfaces and a glass chemist intrigued with electrical conductivity in glass. The original experiment that led to the sodium-sulfur battery might have taken place almost anywhere; it was, in fact, an intriguing and clever experiment in which the investigator was looking for stressed-induced diffusion effects in soda-bearing glass based on the crystallographic identification of the positions of the sodium ions in the quasi-lattice. He found the effect he was looking for but, incidentally, also discovered some rather remarkable electrical properties of the glass associated with the high mobility of the sodium ions. In seeking a potential use for this remarkable property he and a co-worker hit upon the idea of using this sodium conducting material as a solid electrolyte in a secondary cell. Guided by a newly acquired understanding of atom movements in such solids and armed with the readily available techniques and talents of crystal chemistry, solid-state physics, crystallography, magnetic resonance, and phase analysis, they proceeded to "crystal engineer" a completely new material with the desired properties and brought home a working battery.

What is meaningful in the context of today's discussion is the sequence of events that led from a random scientific observation to a useful device that has caused a major industrial enterprise to take a bold new look at possible alternatives to its basic product and assume an optimistic view toward the potential of electrical propulsion. The same experiment might have been done in any number of laboratories. But the circumstance that took place in a laboratory in which there was a built-in sensitivity to the need for new energy storage devices undoubtedly catalyzed the battery into being rather than, say, a thermoelectric device or any number of alternative uses of the remarkable new material we invented. Furthermore, it was necessary to deploy a multitude of available disciplines each of whose practitioners saw in a battery a substantive substitute for his normal life's blood—publication.

Before I leave the subject of pollution, I would dwell briefly on another aspect of our involvement that further illustrates these environmental factors that I consider so essential to the utilitarian deployment of science. Since the discovery of the laser, countless laboratories have sought ways to realize the bountiful potential that had been projected for it ever since its discovery. So far the only ones making money on lasers are those who are selling them to others who are trying to find ways of making money with them. We have a laser group, too. They are a most reputable bunch. Some of them are known to many of you. Their preoccupation has been more with the nonlinear optics that can be studied using lasers as optical sources than with the lasers themselves. This group was the first to recognize that an intense electric field can be produced by focusing a laser, thereby producing breakdown in gases. The laser spark thus conceived has since been used in conjunction with time-of-flight mass spectrometry (a Goudsmit, not a Ford, invention) to study for the first time the fast reaction kinetics of combustion. While I cannot yet report to you any great discovery or invention resulting from these experiments, as a scientist I cannot help but feel that this type of experiment offers the greatest potential for cleaning up the combustion products of gasoline. As a poker player, I cannot help but feel that the rewards to our industry and to mankind justify extensive betting on this type of research, particularly in an environment that is prepared to seek its ultimate exploitation.

I would like to return now to the field of transportation. Wherein, you might ask, lies the scientific component of transportation. Unlike, say, communications or ordnance, there are no obvious underlying disciplines which can combine to advance well-defined frontiers. Because of our intense preoccupation with problems of transportation—we are in the transportation business not just the automobile business—we asked ourselves the same question. To find some answers, we put together a small group—a physicist turned operations researcher, an electrical engineer, and others—to see what dimensions can be used to describe and understand transportation. We have learned a few interesting things, one of which I should like to describe for you.

I believe that we all recognize intuitively that transportation affects the way in which a region develops: that transportation is a part of the urban system incapable of isolated analysis. Further complicating this analytical problem is the fact that we are confronted with a profusion of technology—hardware ranging from conveyor systems to VSTOL aircraft—with little quantitative knowledge to guide its application.

One of the first questions which we asked ourselves was: are there any mathematical tools or analogies which would enable one to quantify and optimize transportation relationships—somewhat in the manner of the use of Monte Carlo methods in the analysis of communications networks. Clearly, the philosophy of systems analysis was indicated, and the lack of an exact solution led to consideration of mathematical modeling and simulation as a means of analysis.

Consider the question of the growth of a city under the influence of population, economics, industry, transportation, topography, and other constraints on land use. It is possible to develop a mathematical model which, based on projections of basic industrial activity, will "move" a region through time, locating new households and commercial activity in accordance with a set of rules concerning accessibility and other factors. We have developed such models and tested them in real world-cities by running them backwards. Once calibrated, such a model can assist the planner in determining the consequences of a decision on, for example, the location of a transportation link. A serious communication problem arises, however, when the planner is faced with a computer output consisting of thousands of population figures in tabular form, and when he must translate his proposed plans into a complicated coordinate system for computer input.

We have made a small stride in the direction of better planner-simulation communications with the cooperation of the Coordinated Sciences Laboratory of the University of Illinois, whose facilities we are dedicating today. A film was generated here that illustrates a hypothetical region which is to be served by a new road. The planner can study the effect of alternate road locations by drawing the proposed route on the face of a scope with a light pen and watching the results of the simulation as they are plotted by the computer.

I would be hesitant to attempt to extrapolate these relatively limited experiences to the exhaustive solution of the many social ills of the world. It does seem to me, however, that one of the formidable obstacles to be overcome is an institutional one. The science and the scientists that will help solve these problems will still be concentrated largely on academic campuses and will continue to maintain their primary loyalty to their disciplines, their professional journals, and peer groups. Such institutions as the Coordinated Science Laboratory are perhaps a step in the right direction, in that they provide a focus for the articulation of needs that transcend the traditional boundaries of the intellectual marketplace. Weinberg sees in this new and urgent social demand a possible role for the National Laboratories. But more than anything else, we need a cadre of scientifically trained accomplished manpower who see a *professional* goal in ministering to the needs of a world that transcends the four walls of their disciplinary prison.

Beyond the needed modifications in the institutionalization of scientific research and facilities lie two additional levels where new approaches are called for: methodology and support.

New scientific methods and techniques have emerged to

help the physical scientist advance his frontiers. Many of these, such as the new solid state electronics and computer technology, were born and bred by military support. The application of science to social problems requires an exploration for useful methodologies that may already exist or may lie hidden in the rich storehouse of emerging knowledge from the behemoth of presently supported research. One technique, systems analysis, which is showing promise in attacking these complex social problems, is also a direct result of a wartime need. Formulation and quantification of goals—often conflicting, occasionally contradictory—and optimum deployment of limited resources to approach these goals is as much a concern of the city planner or public health officer as it is the concern of the military commander. Methods developed by mathematicians for evaluation of weapons systems effectiveness may, for example, provide the only practical way of comparing the probable consequences of alternative solutions to social problems.

Finally, the scientific community and the government community must reexamine the support base for research. Nearly all present-day basic science derives its support from the military and space establishments, from the AEC and from the National Science Foundation. The scientific community must now face squarely the challenge to press other segments of the government and society to support science if they are to expect to derive benefit from it. The challenge to government is equally sobering. Those sectors of government more directly concerned with the problems of social change, e.g., HUD, DOT, HEW, Commerce, must accept a share of the responsibility for the support of science. The very arguments advanced to create the military support agencies for basic and academic science are just as valid in arguing for support of science by these other agencies. If they expect the scientific community to generate interest and concern for the problem lying within their purview, they must provide the principal catalyst for coupling science and scientists to their needs. Financial support is the best catalyst I know for generating this coupling. Maybe what I'm asking for is a social ARPA or an ONR for HUD and an OSR for DOT. However it is accomplished, the precedent has been established by these foresighted mission-oriented military agencies. It remains for the nonmilitary ones to profit from the national experience.

REFERENCE
1. President's Science Advisory Committee, "Restoring the Quality of Our Environment" (Environmental Pollution Panel), 1965.

3   CHALMERS W. SHERWIN

The Coupling
of the Scientific and
Engineering Communities
to Public Goals

Scientists and engineers in universities face a rapidly chang-ing world. These changes may be characterized by seven basic problems which I will state as briefly and realistically as possible. Following the statement of the problems I will describe a proposed four-part solution.

The Problems

1. The Size of the Research and Development
   Establishment

The Federal support for R & D is now about 17 billion dollars per year, and private support another 4 or 5 billion dollars per year. This is about 3 per cent of the G.N.P., up a factor of more than 10 in the past 2 decades, and it will un-doubtedly continue to rise.

Chalmers W. Sherwin: Independent consultant and leader of a special task force reporting to the Director, Office of Science and Technology, Executive Office of the President, Washington, D.C.

2. High Unit Cost

Currently, if all costs are realistically calculated, including all support and overhead, it now requires about $50,000 to provide one professional man-year of research or development effort. This is $200 a day, or $25 an hour. Such an expenditure rate requires a convincing defense, particularly when supported by public money.

3. The Dominance and Nature of Public Support

About 90 per cent of the research costs and 66 per cent of the development costs are now provided from public funds, almost all federal. Even as recently as 3 decades ago, when I was a graduate student, research was mainly supported by philanthropic funds and engineering mainly by industrial funds. The values and mores of any society change slowly and it is not surprising that the dominance of public support has not been fully appreciated.

Most important to realize is the fact that the public, and the Congress which represents it, is mission oriented, or, more accurately, social problem oriented. In thinking about managing and funding public effort, Congress and the federal agencies almost invariably regard both research and development as just one more tool to help them do their job better.

Quite naturally, the public problems with the greatest sense of urgency get the most R & D support. Thus, for example, one can argue that about 90 per cent of all federal R & D is funded out of a deep and urgent sense of concern or "fear"—mainly of potential foreign enemies and disease. In this manner one can account for all the R & D budget of the Department of Defense, much of the Atomic Energy Commission and much of the National Aeronautics and Space Administration (clearly motivated by a concern for national security with respect to space) and all of the budget for health research in the Department of Health, Education and Welfare. Next in efficacy is the motive of economic advantage or essential public service, such as natural resource development, better roads, improved weather prediction, improved education, etc. These R & D budgets add up to about 8 per cent total. Finally we have about 2 per cent (primarily in the National Science Foundation) which is research supported without reference to specific, short-term public need.

To expect that the "distribution of motivation" of the federal R & D budget of:

90% fear of enemies, or disease
8% public economic need
2% science, as a nonspecific public investment

will significantly change in the next decade is unrealistic. I believe that, even though we are really a very rich nation (on the average), the pressure for available public funds for at least the immediate future will be so great that R & D will continue to be made available in substantial amounts only to meet urgent social needs that are understandable and meaningful to the average citizen.

4. The Payoff Time Is Long

It is generally agreed that, on the average, engineering development and technology efforts in the public sector start becoming effective about five years from initiation.

There is, I believe, good evidence that the payoff time for applied research (that is, the time from initiation to the appearance of useful end-items employing the results) tends to be about 10 years.[1]

Finally—and this is perhaps more controversial—the payoff time for undirected research tends to be 20 or more years. This at least is one estimate based on military R & D.[1]

5. The Payoff Ratios of the Stages of the R & D Process Are Not Well Understood

Many economists are beginning to believe that deliberate R & D effort is the primary source of the increase in the intrinsic efficiency (i.e., in the output per man hour) of workers of all types.[2] Also, for many years the allocators of public funds have generally believed the overall payoff on R & D is favorable. The problem lies therefore not in the defense of the overall payoff, but rather in the rational determination in the balance of the various stages. To get such a rational balance one needs to know the payoff ratios and characteristic delay times for the different types of effort: engineering development, technology, applied research, and undirected research. This knowledge should permit an objective, economic allocation of public efforts in the R & D sequence. The conventional wisdom has it that the payoff ratio increases much more rapidly than the discounted investment as one moves toward the research end of the spectrum. This is probably true but needs convincing proof. Even with proof, long delayed payoff times are a problem, for, as Walter Lippman has observed: "I have been asking myself why a country which is as rich as we are today should feel itself compelled to economize at the expense of its children and of its poor. There exists, I have come to think, some kind of rule which in a democratic society limits what the voters will stand for in the way of sacrifice for the public good—the public good which is not immediately, obviously, and directly to their own personal advantage." [3]

6. Universities Are Discipline Oriented

By tradition and probably for sound educational reasons, the basic structure of the university is discipline oriented. Certainly, up until now, it is the disciplinary departments which have given the soundest education, the most effective research tools, and the most rapid advance of basic scientific knowledge. The problem, however, is to provide the "impedance match" between the discipline-oriented university and the mission-oriented society which surrounds it and supports it. The classic analysis of this problem has been given by Alvin Weinberg.[4] One is hard pressed to add anything new to his penetrating observations.

Although this situation has been obvious for some time, the typical university still seems to want to turn out discipline-trained students and to perform discipline-oriented research. Society, on the other hand, wants increasing numbers of graduates who are interested in and can directly attack social problems. These are almost invariably multidisciplinary. Society also wants research projects performed on these same multidisciplinary social problems and frequently turns to the universities only to find their excellent research capability ordered neatly along a disciplinary structure—even in applied science and engineering.

7. The Large Accumulation of Scientific and Technical Knowledge

The rate of production of new fragments of knowledge —represented by "significant" individual papers and reports —now exceeds one million per year. Concern has been expressed as to whether this enormous output is being efficiently used either to advance science itself or in practical affairs. One thing is known: A number of different contributions usually need to be related to each other in some manner if they are to be effectively utilized. The number of paired associations increases as $N(N-1)/2$, and the number of triple associations increases as $N(N-1)(N-2)/6$, etc. Thus as N increases, the "relationship problem" rises much more sharply. It is hard enough just to keep an orderly and usable record of the existence and essential content of the individual contributions, not to mention their relative quality or the significance of their potential interrelationships.

Finally, there is the "translation problem" of interpreting and making more understandable an increasingly sophisticated science to the practical user. These then are seven problems associated with what might be called the maturing of the R & D enterprise, particularly as it relates to the support of public goals.

I will now turn to four specific actions which universities can take, or participate in, which should contribute to the solutions of the seven problems.

Some Proposed Solutions

1. The Universities Should Promote Institutional Support

The universities should base their claims for increased federal support for institutional and "block" or area funding in science and technology on their unique educational mission. No matter where they are located geographically, they are serving a national "market" for highly educated scientists and engineers. Federal support is needed and is appropriate. The short-term payoff is in the educated people produced. The basic disciplinary framework of the university is preserved by this type of funding. Undirected science and longer range applied research are supported and justified as an essential part of graduate education, rather than by their applications. I would suggest that the science and engineering graduate education activities need about 50 per cent of their total support to be of this type. I believe that thoughtfully developed claims for this kind of funding are both reasonable and politically acceptable, and will lead to more stable funding and strengthened institutions.

2. The Universities Should Promote and Operate Problem-Oriented Laboratories or Programs

Weinberg points out clearly that since the students are going to end up in the real world working on multidisciplinary problems, it is perhaps not a bad idea to have them gain some experience in graduate school. Also, as we have noted, no matter how some might wish it to be otherwise, this is where the big research dollars are found. The objective therefore is not to fight reality but to exploit it.

The proposed solution is for each university to set up an array of problem-oriented, and therefore, almost invariably, multidisciplinary laboratories. These laboratories will seek a balance between relevant basic research (which can be treated almost as an overhead charge), applied research, and engineering efforts. Locally selected program directors will come from either the science or engineering side of the house, but in either case must be well balanced and capable of understanding and appreciating both. The staff working in the programs will usually have a regular home in a disciplinary department, and will move into a program on some sort of rotation or part-time basis. Graduate students in the program will do theses in both the interdisciplinary areas and in the relevant disciplinary areas supported by the program. The boundary lines between basic and applied research and between applied research and engineering will be ill defined. A reasonable balance might be one-third generally relevant basic research and two-thirds applied research and engineering.

To usefully attack most problems and have adequate internal flexibility to adjust the effort to the needs, a program requires a certain minimum level of effort. In most areas this minimum is about $200,000 a year, but often $1 million a year is needed for a really effective program effort.

Assuming that an adequate base of institutional support is maintained, programs have many advantages to the universities: (a) they draw the loyalty of the staff away from Washington and toward the university, (b) they develop administrative skill in research management within the university, (c) they draw science and engineering together, (d) they produce graduates who, though having a basic disciplinary education, are interested in real-world problems and are capable of team effort in their solution, (e) they reduce the contract administration load (compared to many individual projects) and, (f) to the degree they are responsive to public needs, they obtain adequate and unusually stable funding.

To the mission-oriented government agency: (a) The program is easier to explain and defend to Congress due to the problem orientation of the effort plus the relatively rapid appearance of at least some practical results. The support of the basic research also benefits from its association with an easily understood mission function. (b) The program develops R & D managers who really understand and become involved in the broader problems of the agencies. They are unusually valuable as consultants and are a primary source of recruiting for periods of service in the government. (One has only to examine, for example, the roster of high level Department of Defense scientists who came from a university for a tour in government.) (c) The successful program, depending as it does on good local management, requires less federal in-house technical management effort than does the equivalent effort in specific project contracts.

Against these advantages, there are certain disadvantages to the government. How does one handle the program which either has or has developed poor local leadership? How does one handle the program which has an obsolete problem of mission? Nonetheless, I believe that the wise and careful expansion of the program type of federal support will lead to both better universities and a more efficient use of government R & D resources.

One word of caution is needed. Institutional support and program support will never cover *all* the research needs and opportunities. There will always be a place for small projects and for research efforts in areas not related to a mission-oriented program. Also, there will be research efforts too large to fund under institutional support. Such activities will have to be justified simply on their long range importance to the public interest.

3. Engineers and Scientists Must Develop Skill in the Economic Analysis of Their Own Functions

To paraphrase a famous statesman, "The economic analysis of science is too serious a business to be left to the economists." The sooner the technical community realizes this, the better. If the era of the 40's and 50's was the era of the scientists in Washington, the 60's is the era of the economist.

At the present level of support demanded by science and engineering, there is simply no alternative to the development of trustworthy means to calculate benefit/cost ratios, particularly with respect to public benefits. There is an essential scientific and technical input needed to understand the way in which scientific and technical efforts finally produce practical results. It is, I believe, a priority task of the technical community to trace out and to understand this process as it has actually occurred. Only then can one rationally forecast the future.

I suggest that all R & D organizations should invest say one per cent of their professional technical manpower on a task assignment or rotation basis working on the problem of understanding the way in which science and technology are utilized. They need to cooperate with an economist in this study. I do not know of any investment of effort more likely to improve the overall efficiency of the R & D operation than this one. Clearly, the overall R & D process as now practiced has a high payoff. It is most likely that developing a systematic understanding of the process will further increase its effectiveness.

4. The Evaluation, Organization, and Compression of Knowledge

Although the specialized research workers with their "invisible colleges" do not appear to have an information management problem, this is not true of most people concerned with catching-up or with using knowledge.

The most practical solution here is for the scientists and engineers to expand their efforts in the systematic evaluation and compression of knowledge. S. Goudsmit[5] has described the need and the process very well. The Standard Reference Data System managed by the National Bureau of Standards is one embodiment of this important process.

Although it costs 20 to 40 thousand dollars to produce the typical paper, experience shows that it costs only about 100 to 200 dollars to have a highly qualified scientist or engineer evaluate it, relate it to other work, and produce a useful "compressed" output. This critical evaluation process could be done for all of science and engineering simply by deflecting about one per cent of the present research manpower into this task. An evaluation center can process 1,000 to 3,000 papers per year using the part-time services of skilled, active research professionals and cost 100 to 300 thousand a year. Such centers must be located at an active research institution. The universities should play a big role in this area, for it is directly related to their knowledge organizing and teaching function.

Conclusion

If institutional support and program support is increased to universities so as to provide the majority of federal support, and if one per cent of the technical manpower is focused on understanding the R & D process, and another one per cent on the evaluation and compression of knowledge, most of the seven problems listed above would be on the way toward being solved.

REFERENCES

1. Sherwin, C. W. and Isenson, R. S., *Science,* June 23, 1967, p. 1571.
2. See, for example, Domar, E. D. *et al., Rev. of Economics and Statistics,* February, 1964.
3. Lippmann, W. "Hard Choices at the Ranch," *Washington Post,* January, 1966.
4. Weinberg, A., "But Is the Teacher Also a Citizen," *Science* 149 (August 6, 1965).
5. Goudsmit, S., *Physics Today,* September, 1966.

4 WILLIAM J. PRICE

The Key Role of a Mission-Oriented Agency's Scientific Research Activities

I am very pleased to participate in this symposium on the interaction of science and technology. I consider the subject to be both important and timely.

Clearly the continual expansion in knowledge and its proper utilization are matters of utmost importance. Currently we are in a period of the most rapid change in history, which in a large degree is both caused and supported by this expansion. However, our understanding of the processes by which science and technology interact to affect this expansion is not commensurate with its importance to society.

The topic is timely, especially because of its special relevance to the current debate on science policy. Discussions leading to improved understanding of science-technology in-

William J. Price: Federal Executive Fellow at The Brookings Institution, on sabbatical leave from position of Executive Director, Air Force Office of Scientific Research. The opinions expressed in this paper are those of the author and do not purport to represent the views of the United States Air Force or of the staff members, officers, or trustees of The Brookings Institution.

teractions can help optimize decisions on the nature and amount of both science and technology required to best serve society.

Scientific research is typically packaged in terms of scientific disciplines, while society's problems almost always appear in other forms.[1] It is generally recognized that engineers and other technologists must always play a primary role in the required communication. The role of the scientific community is not recognized or understood nearly as fully.

The idea that "good" science must be "pure" academic research where the concept of purity implies conscious disengagement from utility is an indication of the need for further understanding. Even though this disengagement is neither necessary nor nearly as widespread as many seem to believe (see for example Medawar's excellent discussion[2] which helps remove this myth), the fact that the idea is brought up frequently underlines the need for additional attention to the scientist's part of the dialogue with technology.

The science-oriented activity in a mission-oriented organization has an important role to play in assuring good communication, inasmuch as it can help optimize the scientist's part of this dialogue. This is a central role of AFOSR for the Air Force, of similar science-oriented activities in other mission-oriented agencies, and of corporate research laboratories for large industrial corporations. This role and the other important functions of science-oriented activities in support of mission-oriented organizations is discussed in a growing body of recent literature.[3-9]

The importance generally attached to the support of scientific research by mission-oriented agencies is underlined by the fact that about 85 per cent of all federal funds supporting university research originate with these agencies. There are, of course, great strengths in the current methods of federal funding and many seek to maintain similar patterns in the future, including the feature that the bulk of the university funds accrue from mission-oriented agencies. It is particularly interesting to note that Sir Gordon Sutherland,[10] recognizing the great value of the U.S. system both to the agencies and the universities, has suggested that Britain adopt the system of university support by mission-oriented agencies.

Some persons have raised concerns about the extensive support of university research by mission-oriented agencies. Handler[11] sees a basic incompatibility between university purposes and the requirements for success in mission-oriented research. Therefore he proposes that the bulk of the funds for academic research come from NSF and NIH. DuBridge,[12] although he sees nothing basically wrong with the whole picture of U.S. support of science in recent years (including the fact that most of the federal funds for that purpose accrue from government expenditures aimed at other national goals), calls for an increase in NSF vis-a-vis the mission-oriented agencies because of the current prevalence of short-range pressures on mission-oriented agencies which tend to be incompatible with the support of university research. Those who quoted [13] the first interim report of Project Hindsight concerning the small contribution of university research to weapons system development, without critical commentary with regard to the applicability of the Hindsight methodology for evaluating the contribution of such research, in effect have supported concerns about the propriety of the support of university research by mission-oriented agencies.

I believe that the concerns described above and other similar concerns are not valid when the scientific research activities of mission-oriented agencies serve their proper roles —the roles which have been generally served in the past and which should be nourished in the future. In this paper, I seek to present further information about these research activities to help assure that they are properly evaluated in the current debates. This is a particularly fitting forum from which to attempt this because the interaction of science and technology comes into the discussion in such a central way.

The following discussion draws extensively on Air Force research experience; further, it deals primarily with the university support program. Notwithstanding, I believe that many of the observations which I make apply equally well to the research programs of other mission-oriented agencies and at least in some respects to federal support of scientific research in industry and government laboratories.

AFOSR Organizational and Historical Considerations

The Air Force Office of Scientific Research is a part of the Office of Aerospace Research, a separate operating command of the Air Force with overall responsibility for the Air Force's corporate research activity. OAR has a budget of approximately $90 million annually for research. AFOSR, with a budget of approximately $40 million, is responsible for a research program that is conducted by contracts and grants. Other major activities of OAR are the Air Force Cambridge Research Laboratories and the Aerospace Research Laboratories, both in-house laboratories with associated contract programs.

The systems development responsibilities in the Air Force rest in the Air Force Systems Command. AFSC has an annual budget of over $3 billion for research, development, engineering, and testing. This organization conducts a great variety of applied research, exploratory development, advanced technology, and systems engineering programs.

The AFOSR program includes about 1,000 separate research investigations at about 200 universities, industries, and other research organizations. The research is selected for support from unsolicited proposals. The selection is made by

AFOSR program managers on the basis of the suitability of the proposed research to the program for which they are responsible and the scientific quality of the work. Research is supported in chemistry, mathematics, electronics, solid mechanics, aeromechanics, energy conversion, general physics, solid state physics, astronomy-astrophysics, and the behavorial, biological, and information sciences. Interdisciplinary fields are also supported as the need arises.

AFSC, especially through its Research and Technology Division, is also concerned with science, particularly applied science. In fact, the amount of money which AFSC spends in the nation's colleges and universities to broadly support technology objectives is somewhat larger than the OAR expenditure to perform the Air Force corporate research function.

The concept leading to AFOSR was set forth in 1949 by the Ridenour Report, a study by a special committee of the Air Force Scientific Advisory Board. The study pointed out that Air Force research and development could not be maintained at the highest level of competence without being closely associated with the general research efforts of the nation's universities. To accomplish this association it recommended that a fraction of the R & D budget of the Air Force should be consistently assigned to contracts with educational institutions for research in broad general fields on problems which, without being directed toward definite applications, are of definite interest to the Air Force. In 1955 AFOSR was established as a separate operating entity, having functioned as a staff agency under the name Office of Scientific Research (OSR) for the previous four years.

Recently we have devoted substantial effort to historical-type studies of AFOSR, including a survey of a group of previously supported AFOSR principal investigators and other knowledgeable persons to obtain additional information on the utilization of results of the AFOSR research programs.

These studies have been worthwhile inasmuch as they have provided us a large increase in specific information showing how the Air Force has benefited from the AFOSR program. Perhaps even more important, these studies have increased our knowledge of the interaction between science and technology and concurrently have aided us in bringing our role into sharper focus.

We are impressed by both the large number and great variety of the AFOSR accomplishments that can be identified as important contributions to the Air Force. Our success in finding utilization through this study shows us that a large amount of additional specifics could be accumulated if necessary.

I shall only touch on some of the highlights of AFOSR accomplishments as a background for the following discussion. Another publication, just released, includes a detailed account of these accomplishments, along with a discussion of the current program and the ongoing role of AFOSR.[14]

We find that the AFOSR has helped colonize many important scientific areas which have turned out to have special relevance to the Air Force, inasmuch as they are generally recognized as underlying important Air Force applications. Colonizing may be described as increasing the chance of important discovery in an area by "raising the temperature" of the world's scientific activity in that field. Through judicious support of phenomena-oriented research and other activities such as symposia, the Air Force research support, amplified by that supported by non-Air Force funds, has affected very significantly the rate of development of important scientific areas—hypersonic phenomena, including hypersonic facilities, magnetic resonance spectroscopy, optimum control theory, visual perception, mass transfer cooling, information theory and many others.

We also find that AFOSR is playing an important role in technical education. At any given time our research program is providing at least partial support for the doctoral research of more than 1,000 graduate students. The overall importance of this support is quite substantial but hard to measure; however, it can be appreciated by recognizing that these students are among the top strata of the nation's graduate students and they are receiving their education in areas particularly relevant to the Department of Defense. Many have gone on to work in Air Force contractor or in-house activities, equipped with knowledge and skills particularly pertinent to their work, because of the previous Air Force association.

We also find that we can identify many specific examples where AFOSR supported phenomena-oriented research has provided important support of Air Force weapons acquisition programs at all phases of the research, development, and engineering cycle. We find this input through new or improved manufacturing techniques, design techniques, instrumentation, and weapons systems component concepts, to mention a few cases. The MAB study[15] on research-engineering interaction has also noted a similar diversity in the types of important interactions occurring between science and utilization. The diverse nature of the interactions are also brought out by Morton's model of the innovative process as a complex feedback-type system,[16] and Piore's presentation at this symposium.[17]

We also find that many scientists supported by AFOSR are consulting for the Department of Defense contractor and in-house research and development activities. In a very real sense, AFOSR support helps these persons achieve and maintain their expertise while they contribute direct practical help to the Department of Defense.

Finally, it is important to note that the AFOSR program has provided research support for scientists who are among the leaders of their respective disciplines (see, for example, the current program listing[14] and the indication from the evaluation of AFOSR research using citation indexing[14] that the research results have often been among the most important in their fields).

Interaction of Science and Technology and the Role of AFOSR

We find that it is helpful, both in describing the role of AFOSR and in discussing the interaction between science and technology,[18] to divide research and development activity into two broad categories—phenomena-oriented science and technology, as illustrated in Figure 1. In technology creative efforts are primarily concerned with synthesis, that is, integration of previously existing knowledge components into operational capability—for example, systems, devices, processes, methods, and materials. In contrast, phenomena-oriented science is more heavily concerned with the origins of the knowledge components themselves.

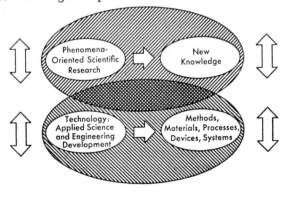

FIGURE 1

Notice that both applied science and engineering development are classified as parts of technology. Applied science is so named because its goal is some sort of application of scientific principles. Thus the name comes from the goal. A phenomena-oriented scientist concerns himself with the elucidation of natural phenomena. Thus his goal is the study of phenomena, and consequently, it is reasonable to call his activity "phenomena-oriented science."

As new phenomena are understood, this new knowledge is made available to the scientific and technological communities in many ways. However, it is important to note that the new information becomes known by the peer group in the world scientific community much sooner than it is known by other groups, particularly those associated primarily with technology utilization. Important new knowledge, such as that being accumulated in the scientific fields that AFOSR is helping to colonize, is known to members of the "invisible college," that is, those researchers active in the particular segment of the research front, well in advance of any formal written publication. Thus new science may forge ahead, relatively independent of an ambient technology.

Similarly, technology usually feeds upon technology in the presence of an ambient science. It has become increasingly clear, especially to the historians of science, that technology events are usually initiated within technology. This means that usually it is difficult to establish a unique correlation between a technology advance and one within phenomena-oriented science. One well-known exception is nuclear power, and its origin in the discovery of nuclear fission. Our study shows that one does not usually find a phenomena-oriented research *result* producing a new and unexpected opportunity which then *stimulates* a new engineering opportunity. This observation is consistent with those of other recent studies on science and technology interaction.[15-21] Instead, we find phenomena-oriented research supporting technology in many other important ways, and thus in a real sense making the advances possible.

Thus the gross picture is that technology usually feeds

upon technology and phenomena-oriented science usually feeds upon phenomena-oriented science. However, at the same time we find that there is a strong, important interaction (almost a symbiosis) between the two spheres of activity as symbolized by the overlap in the figure. When one looks deeper[14-28] one finds many possible important avenues of interplay between science and technology. Some of these are actively utilized; others need to be further developed.

The nature of this interface is dynamic, varying greatly among different science-technology pairs and with time for a given technology. Industries that are involved in communications, computer technology, and modern instrumentation are much more closely coupled to science than are the railroad and agricultural equipment industries, for example, and transistor technology was much more closely coupled to solid state physics 15 years ago than it is today.

Further indication of the changing nature of the interface is the suggestion[22] that even though historical studies show two spheres of activity, the present-day boundaries between science and technology are becoming blurred in many areas of endeavor.

It is found that interactions leading to utilization of phenomena-oriented research are usually initiated by persons who, having the urgent need for knowledge, search for the solution through prior research. The scientists who are consulted also play a very key role on their side of the dialogue, providing knowledge and interpretation from their field.

The fact that new knowledge, originating in phenomena-oriented research, often has implicit in it important new opportunities for exploitation suggests that great advantages, particularly in timing, can be realized when these opportunities can be recognized on the research side. This appears to be an important area for increased attention by phenomena-oriented research activities toward the end that the initiative can be successfully taken by the scientist more frequently, even though there is some serious question of the value of a solution looking for a problem in the overall scheme of things.

Our conclusion from these studies is that the conventional picture, which emphasizes a process with unique scientific events being followed in an orderly manner by applied research, development, etc., is usually not borne out. Since that picture appears to be the exception rather than the rule, it is misleading to attempt to elucidate the contribution of phenomena-oriented science by studying this process primarily. In fact, the failure to observe a large number of such cases could lead to a nonobjective backlash in which the *real* (and very important) process involving the flow of an immense number of items of information across the technology-science interface is not recognized.

Any study, for example, of a series of weapons systems, aimed at isolating the points of origin by identifying key events, will often reveal them to lie within technology.[20] A dangerous conclusion which might be drawn from such results is that science (especially current science) is of little help in the development of weapon systems. Actually, what our studies show is that current science is exceedingly important to the development process, not so much from the point of view of origin, but rather in continuously providing many useful items of information, and particularly in providing an ever growing sophistication in handling technological problems. This sophistication is made possible by the increased understanding of the phenomena involved and the increased educational level of scientists and engineers. It is this continuous support which should be the primary object of study rather than the question of origin, when the impact of scientific research is to be evaluated. Recent literature contains numerous comments on the impact of scientific research which generally support this point of view.[17,21–28]

Similarly, it would be a mistake to try to have a large fraction of the Department of Defense phenomena-oriented research programs arranged to support specified technology goals, such as the exploratory development projects of the Department of Defense. This type of programming will not be successful since it assumes direct, rather simple relationships between the science program and the technology. For example, if the Air Force were to use primarily this type of programming looking back from technology needs toward science, it would end up with more applied science (or technology, in terms of this model). Although this type of activity would be very valuable, it would simply be adding to the excellent applied research programs already being conducted by the Research and Technology Division and others, and AFOSR would not be performing its assigned mission. Rather, the role of AFOSR is to capitalize on its strong identity with phenomena-oriented research to bring a new type of capability to the Air Force research and development activities, thus complementing the many excellent technology activities.

This knowledge of the interaction between science and technology provides guidance for the mission emphasis of AFOSR. AFOSR can be visualized as an activity which, because of its intimate involvement with both the scientific community and the Air Force, can help provide an effective interface between these two communities. AFOSR can attract the interest of the world's top scientific talent, since it is a science-oriented organization with a well-established reputation in the scientific community as a good research agency with which to work. At the same time the AFOSR staff members have the organizational position, and a growing body of experience and techniques for carrying on an effective dialogue with the Air Force technology community.

In providing this interface, AFOSR engages in two types of activities. First, it supports high-quality scientific research chosen because of the particular interests of the Air Force. The selection of these areas may be motivated either by seeking to pioneer new fields of science holding out high promise for generating the new knowledge from which new technologies or new operational possibilities may evolve or it may be motivated by helping various development or other user groups solve certain difficult classes of important problems by providing a fuller understanding of the phenomena behind them. Second, it helps provide communication between the scientific community and the Air Force. This is a two-way communication—needs to the research program and scientific information to the user. The AFOSR project scientists play the key role in this communication or coupling activity. In addition, part of what we purchase through contracts and grants is primarily designed to provide communication. This part refers not only to the symposia we sponsor, but to the connecting-type research which allows us to keep abreast of a variety of scientific areas largely supported by other agencies, but nevertheless important to the Air Force because of rapidly emerging scientific developments.

Specific Areas for Improved Interaction Between Science and Technology

The preceding material provides the context for highlighting areas of improved interaction between science and technology. In summary, my proposition is that, for many important aspects of the interaction of science and technology, it is essential to recognize that there are two communities—phenomena-oriented science and technology—which, although they interact with each other in many effective ways, nevertheless have separate identities. Further, the existence of these two communities leads to a key role and special challenge for organizations such as AFOSR in bringing about improved interaction.

I believe that there is considerable potential interest within each community for improved interaction. However, an improved dialogue can often be cultivated best by recognizing the needs and strengths of each community. It is particularly important to build a growing dialogue that recognizes the individuality of each partner.

It is inevitable that these communities will often be organizationally and geographically separated. Further, they usually will lack active emotional identity with a common purpose. Nevertheless, there are many situations where the merging of the two communities by organizational fiat would have a dampening effect on communication rather than the intended improvement. The continued existence of a corporate research function, OAR, may be considered as concrete evidence that the Air Force recognizes the importance of a viable relationship between the scientific and technology communities which allows for the continuing strength and identity of each. OAR provides bridges between the two communities without perturbing unduly the nature of either. This does not argue against the importance to the Air Force of organizations such as in-house and industrial research and development laboratories which incorporate a continuous spectrum from operational requirements to fundamental research. Rather, it says that both types of Air Force involvement with science are essential.

AFOSR meets two special challenges of a continuing nature as it helps serve this interface function. One is the choosing of the appropriate areas in which to support research—the planning function. The other is helping improve the dialogue between the scientific and technological communities—the coupling function.

Planning Phenomena-Oriented Research in AFOSR

At AFOSR we are giving a growing amount of attention to the planning process.[29] We believe it is important to do so because the selection of a long-range research program for a mission-oriented organization is a difficult matter for which few guidelines exist, and because the matter of scientific choice is growing rapidly in importance in discussions of national R & D policy.[30,31]

There are, of course, special planning considerations appropriate for AFOSR and those other research agencies that support a mission-oriented organization. On the one hand, a proper planning procedure brings about choices of broad scientific fields, and of specific work efforts within these fields, such that the scientific research program has a "center of gravity of interest" which meets the needs of the agency it supports. At the same time, the distribution must not be restricted by too narrow a definition of relevance. It must be recognized that some areas of science have special importance for several (perhaps all) of the mission-oriented organizations and that support of these areas by more than one research agency can be important both to provide these organizations communication with the fields and to assure that there is adequate support in these vital areas. Further, the distribution must attempt to recognize both the uncertainties in our knowledge of the scientific results which will be obtained and, even more so, the unexpected avenues of utilization of new scientific findings and therefore the unexpected relevance to the mission of the agency.

Inputs appropriate for consideration in planning the AFOSR research program come to us in many ways, both from within the Department of Defense and from the external scientific community. They come through organizational channels and, most important, they come through the many essential activities of the AFOSR program managers. Information about the long range needs of the Air Force comes to us from top management. The Secretary of Defense and of the Air Force, the Chief of Staff of the Air Force, and others provide various internal documents. We receive important guidance from AFSC. Two examples are the AFSC Planning Activity Report and the Technical Objective Document. The latter, for example, sets forth the Air Force interest in each of the thirty-eight technical areas. Inasmuch as we know that it is just as misleading to expect a neat flow of research needs from technology as it is to expect a smooth flow of research results into use, we also look beyond AFSC for requirements. For example, our interests include activities in support of military assistance programs, personnel management and training, logistics, and other needs which we find arising from other parts of the Air Force.

Using these various sources of information, we have developed a list of technology areas to be considered in assessing the relevance of research. The relevance is studied with the aid of a large matrix showing the relationship of the technology list to a similarly detailed list of the scientific areas. The obtaining of the above information is the responsibility of line management and their supporting staffs. In parallel to these activities there is the continual concern for planning at the individual program manager level.

An AFOSR program manager typically has about 20 active contracts and a million dollar annual budget. Each program manager is responsible for one or two subareas; surface physics of solids, control theory, and structural mechanics are typical subareas. Each program manager must engage in a variety of activities—both professional and managerial. As in other similar activities, the qualifications and motivation of the scientific staff are matters of utmost importance.

It is clear that the program manager must have a personal knowledge of the emerging opportunities of science, as discussed later. It is also very important that he have appropriate meaningful personal contacts with those persons throughout the Air Force interested, or most likely to be interested, in the research program for which he is responsible. Here a lot of personal contacts need to be established and maintained through visits, correspondence, special reports, program reviews, participation in joint task groups, etc. A personnel policy which encourages transfers between other parts of the Air Force R & D organization and AFOSR is very helpful in developing properly qualified program managers.

There are several techniques which the individual program managers or groups of program managers have found particularly useful in planning. Let me quickly outline a few of these. We just sponsored a workshop on Fundamental Problems of Future Aerospace Structures in which the Franklin Institute and AFOSR brought together 15 key engineers from diversified aerospace industries to present their views of research needs in structures. The analysis[32] of the material presented at this workshop is an important input to the planning of our solid mechanics program.

During the last several months we have held a seminar on long-range research required to support limited conflict. The primary purpose of this series, which consisted of a wide variety of presentations by speakers having intimate knowledge of limited conflict problems, was to increase the sophistication of AFOSR staff members in selecting appropriate long-range research problems to support that Air Force mission.

In our combustion dynamics program we have an an-

nual meeting of all of our contractors, along with representatives from the Air Force R & D organizations. At this
symposium contractors present their research results and the
Air Force representatives describe problem areas which they
see as having areas of interest to research groups. Another
fruitful technique is in-house advisory committees. Carefully selected members from throughout the Air Force technology community meet on a semiannual or annual basis
with groups of individual program managers.

Other significant examples of special techniques which
I will simply mention are state-of-the-art reviews, participation in ad hoc studies of research and technology utilization
for operational needs, and participation in interagency coordination groups such as the Interagency Chemical Rocket
Propulsion Group made up of research and development program managers.

The third absolutely essential input for our planning is
current information on the continually emerging opportunities for scientific research. Comprehensive studies on the
opportunities and needs of science, such as the series of
subject matter reports prepared under the auspices of the
NAS Committee on Science and Public Policy, are important inputs. Further, our nine research evaluation groups,
one for each of our principal scientific areas, and the Scientific Advisory Group for OAR, which all together include
approximately 100 of the nation's leading scientists, provide
fruitful guidance more specifically tailored for our needs.
We also have special studies of various types, such as the
chemistry study currently being run for us by the NAS–NRC.

The most vital source of planning information is each
AFOSR project scientist's current knowledge of the research area for which he is responsible. He is in an ideal
position to be knowledgeable of the emerging fields of science because he continually receives unsolicited proposals
from scientists seeking support. The originators of the 2,000

formal proposals and several thousand informal proposals which AFOSR receives each year are ready, willing, and able to provide this education to the AFOSR staff. This activity supplements in a very effective way the other professional activities, including sabbaticals, by which the AFOSR project scientists seek to be knowledgeable of the emerging fields of science.

I am involved in a continuous dialogue with the six managers of the principal scientific directorates, my immediate staff office in charge of planning, and other key AFOSR program managers, in order to determine the appropriate balance of our activities. Once a year AFOSR and its counterparts from the other areas of OAR, joins Headquarters OAR to formulate the OAR Five-Year Plan. Also present are selected members from Air Force organizations using research. This Plan is a comprehensive statement of the specific scientific areas in which it is felt research should be supported by OAR. The OAR Plan, which we had helped to formulate, becomes the guidance which we receive from higher headquarters. It is revised annually in order to keep it viable in terms of responding to new scientific opportunities or to improved understanding of future Air Force needs for research.

In attempting to maximize the contributions of our phenomena-oriented research program to the overall mission of the Air Force, we are faced with many choices. We, of course, have no magic formula by which to do our planning and we suspect that none will ever be found. We do believe, though, that searching for answers to these all-important questions in responsible and intelligent ways further optimizes the contribution of our activities. Although scientific excellence is always a prime consideration in what we support, we are bringing the relevance and other considerations into play in many ways and we continue to search out additional means to further improve our effectiveness in this area.

AFOSR Coupling Activities

In our coupling function we seek to help bring new knowledge and understanding from the world of science to Air Force technology and to provide the appropriate feedback of future operational requirements to the scientific community; in other words, we seek to support and improve the dialogue which exists between the scientific community and the technological community responsible for future Air Force capabilities.

Coupling occupies all parts of AFOSR to some degree. The AFOSR directors and myself concern ourselves with methodology, particularly seeking to place management emphasis on those techniques which hold promise for improved effectiveness and efficiency. The individual AFOSR program managers are the primary focal points for coupling, in that they work directly with their counterparts in both the scientific community and other parts of the Air Force. We also maintain an AFOSR Research Communications Office which provides the AFOSR directors and program managers staff support for both the planning and coupling functions. These various coupling activities have been described in detail elsewhere.[33]

We find the coupling function to be both challenging and difficult. Much of the challenge is associated with its open-ended nature, both from the standpoint of the need which seems to exist and the large variety of opportunities open for exploitation. This difficulty is due to the nature of the communication that is required and to the inherent difficulty of improving coupling mechanisms which in many respects are already operating reasonably well. The bringing about of improved coupling usually requires a high degree of professional competence and ingenuity. In coupling, as in planning, it is very important for an AFOSR program manager to develop and keep current contacts with counterparts in the Air Force applied research-exploratory development community. Many of the planning activities which are described above also serve the coupling function so the present discussion can be briefer.

Coupling also requires the AFOSR program manager to keep in close touch with scientific research activities. This latter interface is generally the easiest one to maintain because the AFOSR organization matches well with the subject matter fields in which phenomena-oriented scientific research is usually accomplished, and because we have direct connections with many leaders in the scientific fields through our contracts and grants. It is important to note that this interface provides potential communication not only with the research activities we support, but what is often more important, it can help bring to the Air Force knowledge and understanding from the much broader area of world science with which the scientists supported by AFOSR are in intimate contact.

AFOSR has always sought to encourage the communication methods that develop naturally in the course of professional activities. We have always believed it important to provide liberal allowances for travel to professional meetings and similar activities. We have vigorously supported publication of research results in the open literature (without previous review, I might add); for example, we have devoted quite considerable effort to this issue within the Department of Defense during the last two years, helping keep this important policy on track in the face of very serious counterforces. We have provided financial support to professional journals when this was required to bring these journals to a self-supporting basis in a timely manner. Included were *Applied Mechanics Reviews, Physics of Fluids,* and *Mathematical Reviews.* We have provided help to professional societies including the Society of Industrial and Applied Mathematics and the Military Operations Research Society. We have helped specialized abstracting services, such as *International Aerospace Abstracts* and *Semiconductor Abstracts,* to get started. We support over 50 specialized symposia each year; many of these result in published proceedings.

We have supported the preparation of a large number of books designed to consolidate knowledge in a field and pass it on to others. We have been concerned particularly with those books that are of use to the practicing scientist and engineer, which could not or would not have been written without our initiative and/or support. The AFOSR Directorate of Mathematics alone has supported 81 volumes in the last 12 years. We have also helped pioneer new types of formalized information exchange mechanisms, including the Bioscience Information Exchange (the predecessor of the Science Information Exchange) and the computerized management control data systems in use by the Department of Defense and currently being adopted by other agencies.

Some of our most meaningful coupling activities arise from the personal involvement of the research scientists that receive AFOSR support. The following are a few examples: trips to Air Force installations to perform consulting services; extended stays in Air Force laboratories; membership on ad hoc committees to study feasibility of various exploratory development programs; performance of research for technology-oriented organizations complementing the AFOSR research; state-of-the-art reviews, either oral or written; special purpose symposia which are specifically designed to bring technologists and scientists together; special lecture tours;

performance of feasibility studies on research phenomena to package them in a form more likely to be useful; and direct consultation with the aerospace industries.

Rather frequently I hear it stated or at least implied that university scientists generally insist on doing pure science and to this end resist or at least resent any involvement in society's problems. My four years of experience in the AFOSR coupling program supports a diametrically opposed point of view. I find that the vast majority of scientists find significant satisfaction and stimulation in making these direct contributions to the Department of Defense programs, in addition to the important, although often less direct, inputs which they are making through their research results.

Project and Programmatic Methods of Support

I need to discuss briefly our methods of administering the extramural university contract and grant program with particular reference to programmatic and project methods of support since this bears directly on the topic of this symposium.

Both project and programmatic methods of support have the same objectives; each provides the administrative and managerial vehicle through which research programs of interest to the Air Force are conducted in colleges and universities. The two methods differ in two ways: in the selection of the individual work efforts which make up the program and in the geographical location of the individual work efforts.

In the project method of support the Air Force program manager has the responsibility for selecting the individual projects or work efforts which, when taken together, constitute the Department of Defense research program in the particular program area; in the programmatic support, the selection of the individual work efforts is largely delegated to the university program director. With the programmatic method of support, all of the work efforts in a given program are generally carried out at one university, while when the project method is used, the Department of Defense program in a given area is carried out by separate projects at a number of universities.

In the project method of support, as used by AFOSR, each contract or grant is an integral part of some larger program which is the responsibility of an AFOSR program manager (see previous discussion under planning). The AFOSR program manager selects the individual projects from a large number of unsolicited proposals, basing his selection on the potential contribution of the proposed research to his program objectives, the competence of the principal investigator, the uniqueness and importance of the idea proposed, and the advice from his extramural and intramural advisors and consultants. In turn, the principal investigator has the responsibility and necessary flexibility to deviate from the specific objectives set forth in his proposal if doing so will help maximize the impact of his research. As indicated above, the program manager utilizes a variety of techniques for bringing about effective interaction between the individual principal investigators making up his program and for coupling this program to the needs of the Air Force.

In the programmatic method of support a program manager is designated at the university. He has overall responsibility for the program, including the selection of the individual work efforts which make it up. These university programs may or may not be interdisciplinary. Likewise, they vary in the extent that they include a concern for applied science and technology along with the fundamental science. The local program manager assumes a responsibility for relating his program to the needs of the sponsoring agency (ies), including working with the Department of Defense technology organizations as well as with the responsible AFOSR scientist. The laboratories of the Department of Defense Joint Services Electronics Program (of which our host the Coordinated Science Laboratory is one of the most distinguished) are excellent examples of programmatic funding. Other examples are the ARPA-funded interdisciplinary laboratories for materials research and the new programs currently being implemented under the new Department of Defense university support program—Project THEMIS.

I am going to almost bypass the obvious and important question of how the two methods of support compare from the standpoint of their contribution to the interaction between science and technology. I'm sidestepping this question because it is too important to treat in the inadequate manner which we would be forced to here. The treatment would be inadequate because the question is too complex to treat quickly and also because much of the information required to evaluate the two methods is not yet available.

The following are some of the complex issues involved in this question:

Relative effectiveness of the agency and university program managers in providing the required interfaces with the agency technology.

Comparison of scientific excellence of work efforts supported by the two methods.

Extent to which university programs involve a spectrum of basic and applied work and interdisciplinary approaches.

Flexibility—short term and long term—considered from both the agency and university standpoint.

Relative compatibility of the objectives, as well as the administrative procedures, of the two methods of support with the fundamental purposes of the universities.

Potential for assuring that the universities can make

their unique contribution to the Department of Defense technology-scientific community dialogue as compared to contributions of industrial and Department of Defense in-house R & D organizations.

Currently about one-half of the AFOSR work with educational institutions is administered by the project method, one-third by the programmatic method, and the remainder by group-type research agreements which combine the work of more than one senior investigator at an institution into one flexible instrument.

My personal position on the two methods of funding university work is that we should continue to use both methods, doing everything we can to keep each completely viable, while at the same time continuing to seek to improve the effectiveness of each method in providing an interface between science and technology. We must consider the best interests of both the universities and the Department of Defense, particularly their long-range interests. If at some later date it is clear that the balance in use of the two methods of support should be changed, then we should act accordingly. In view of the lack of such evidence at the present time, current efforts should be resisted to make further large increases in the amount of programmatic funding by discontinuing project-type funding.

Should Applied Science in Universities Be Increased?

We have been asked to consider the improvement of science-technology interaction through increasing the amount of applied science carried on in universities. In dealing with this question, one must first agree on terminology. In this paper I have defined applied science as part of technology—the search for new products, devices, etc. Consequently, my definition points to a negative answer to the question of applied science increase, since clearly technology is not an appropriate activity for *major* concern by a university. On the other hand, if one uses E. Teller's definition of applied science[34] as the activity which bridges "pure" science and engineering, one reaches the opposite conclusion. It is not only appropriate, but it is also important that universities be involved in the dialogue between the phenomena-oriented scientific community and the technology community. The growing AFOSR activity in planning and coupling, already described, characterizes our understanding of the nature of the appropriate dialogue.

Except in the case of the medical profession where university hospitals are major instrumentalities for the utilization of medical science and technology, the universities do not provide self-contained organizations with operational missions requiring the support of a science-based technology. Further, the existence of industry and in-house R & D organizations, which provide very extensive and capable means for utilization in nearly all fields of interest to the Department of Defense, removes both the necessity and the desirability of establishing such organizations within universities. Consequently, even though extensive additions of applied science and technology were made to universities, the interface between university science and its utilization would still largely remain external to the university.

The one great advantage I can see in having some applied science in the universities is that its presence helps catalyze vital science-technology interactions elsewhere. Those university scientists who are primarily knowledge oriented can be aided in their secondary function of seeking to couple science and technology in cooperation with AFOSR or otherwise, by association with colleagues who are doing applied science. For example, I believe that the fact that the Coordinated Science Laboratory on the University of Illinois campus has been concerned over the years, at least in part, with applied work has had an important impact on the concern with utilization in other parts of the University program. Dean Alpert's paper[35] for this symposium presents an excellent discussion of the methods and rationale of the programs by which the University of Illinois continues to pioneer in this important activity.

In summary, I believe that the value of applied science in universities is not primarily in providing local interactions by which growing numbers of university scientists can work on the problems of society, but rather that applied science within a university helps catalyze effective dialogue elsewhere. Toward this end, it is important that the Department of Defense continue to support relatively large amounts of applied research motivated by its technology organizations. Currently over one-half of the Department of Defense monies going to colleges and universities comes from these organizations. These funds certainly should not be allowed to drop as is presently programmed. Nor should it be required to decrease phenomena-oriented university research in order to support applied science, particularly in view of the present squeeze on Department of Defense research funds.

Concerns About Current Trends

Finally, I want to discuss concerns which I have about some current trends in policy and in funding for scientific research. This bears directly on the topic of this symposium. This is a time of many diverse demands on the national budget, and it is no wonder that research budgets and methods for setting priorities for scientific research activities have come under the public spotlight. A redistribution of research support is occurring. Also, there is great concern for utility, shared by most of us at this symposium. There are, however, indicators that we are in danger of serious maladjustments caused by applying over-corrections or un-wise corrections.

In attempting to assure timely utilization of scientific re-search, while simultaneously being forced to present con-vincing arguments for budgets to an increasingly diversified audience, research activities may be tempted or even re-quired to put too much emphasis on such factors as the *promise* of practical advance and geographical distribution. The recent action in which Congress, in taking $12.5 million from scientific research in the fiscal year 1968 Department of Defense appropriations bill, designated that this money should be cut from the university research, but not from Department of Defense Project THEMIS (which has both the geographical distribution feature and a promised rele-vance) is a concrete part of the trend that has been build-ing up for some time.

Figure 2 shows the trends in the support of research, both federal and OAR totals. Since funds available for OAR have not kept up with the increase in the cost of doing re-search (approximately 6 per cent a year), the total amount of research which we are able to support in the nation's col-leges and universities has dropped somewhat during the last six years. Over the same period the fraction of the total federal research that is supported by the Air Force and by the Department of Defense as a whole has dropped rapidly. At the same time, increasingly larger amounts of the funds available to us have been used for new programmatically-funded efforts and for special-purpose research activities. The overall impact of these trends has been a marked de-crease in our capability to maintain a viable interface with the continually emerging new research activities. This is particularly serious as we lose our capability for funding out-standing *young* scientists who are continually appearing in the scientific community.

Over the years beginning with World War II there has

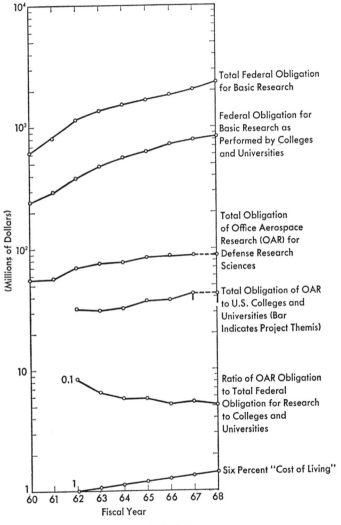

Total Federal Obligation for Basic Research

Federal Obligation for Basic Research as Performed by Colleges and Universities

Total Obligation of Office Aerospace Research (OAR) for Defense Research Sciences

Total Obligation of OAR to U.S. Colleges and Universities (Bar Indicates Project Themis)

Ratio of OAR Obligation to Total Federal Obligation for Research to Colleges and Universities

Six Percent "Cost of Living"

Fiscal Year

FIGURE 2

come into existence an effective and important working relationship between the Department of Defense and the university research community. For the above, and other related reasons, there is considerable cause for concern that this relationship will seriously deteriorate if current trends are allowed to continue.

I believe that it is important to this nation that the effective working relationship between the Department of Defense and the universities be continually strengthened; certainly it should not be allowed to deteriorate. Its importance to the interaction of science and technology supports this point of view. The Department of Defense university research program and that of other mission-oriented agencies is also quite important to the universities from the standpoint of obtaining adequate support of scientific research.

Persons in the academic community and in other quarters often support a plan which proposes to continually shift the funding of university research from mission-oriented agencies to the NSF (many of these same persons often erroneously equate research supported by a mission-oriented agency with applied research; e.g., see Handler's discussion).[11] During the years which this plan has been pursued, the university research support by mission-oriented agencies has tended to level off, but adequate compensatory increases in the NSF program have not been forthcoming.

Certainly Congress and other persons responsible for science policy and funding will continue their critical studies for some time. I feel optimistic that these studies will lead to the provision of adequate funds for university research on a continuing basis. However, I believe that Congress is much more likely to understand and support a program which provides good balance between that for science-dependent mission-oriented organizations and that for NSF than it is to provide large increases for the relatively non-utilitarian NSF program.

Those faced with the problem of developing the overall science policy for the nation have many complex issues with which to deal. It seems clear that one very fruitful avenue to pursue is the development and presentation of the important role of the scientific research activities of mission-oriented agencies.

Summary

In this paper I have discussed the functions of science-oriented organizations in mission-oriented agencies, using AFOSR as an illustrative example. I have described these organizations as having a key role to play in the interaction between science and technology. By making the proper choices of scientific areas for support and by helping to improve the dialogue between the scientific and the technological communities, these organizations serve a function which is very important to both the agencies which have special responsibilities for solving the problems of society and the university-based scientific community which must provide education and new knowledge. The programs of these organizations bring benefits to both the universities and the agencies which cannot be obtained through agency-supported research that is closely related to short range needs nor through the scientific research supported by other agencies.

My proposition is that the best interests of the nation will be served by continuing to provide the bulk of the research funds to universities through the scientific research activities of mission-oriented agencies. Arguments against this can be seen to be largely invalid when they are analyzed in light of the proper functions of these organizations. The mismatch between the problems of the mission-oriented agencies and the disciplinary structure of universities is the central reason these organizations are needed, not a cause for shifting their funds to NSF and NIH as Handler suggests.[11] Furthermore, when the value of university-based research to an agency is considered in the context of a detailed picture of science-technology interaction, there is satisfactory proof of great value received in the past and expected in the future, not the lack of rationale for the agency expenditure of funds that certain Project Hindsight inspired comments have indicated.[13]

When von Karmen was awarded the First National Medal of Science in February, 1963, President Kennedy said, "I know of no one else who so completely represents all areas involved in this medal—science, engineering, and education." von Karmen replied, "I hope that my work has shown that the college professor is of use." Certainly von Karmen's hope was well founded. However, the science-oriented activities in mission-oriented agencies have an important continuing role to serve in order that the continual expansion in knowledge and its proper utilization will work together to provide optimum benefits to society.

REFERENCES

1. Weinberg, A. M., "But Is the Teacher Also a Citizen," *Science* (August 6, 1965), pp. 601–606.
2. Medawar, P. B., "Anglo-Saxon Attitudes," *Encounter* 25:2 (August, 1965).
3. *The Fundamental Research Activity in a Technology-Dependent Organization*, The American University, Washington, D. C., April 26–29, 1965. AFOSR 65-2691, Request #628747, Clearinghouse for Federal Scientific and Technical Information.
4. Goldman, J. E., "Basic Research in Industry," *International Science and Technology* (December, 1964), pp. 38–46.
5. *Planning Phenomena-Oriented Research in a Mission-Oriented Organization*, Seminar at 12th Institute on Research Administration, The American University Center for Technology and Administration, 24–27 April, 1967. AFOSR,

1968; also Quinn, James Brian and Cavanaugh, R. M., "Fundamental Research Can Be Planned," *Harvard Business Review* 42, No. 1 (January–February, 1964), pp. 111–23.

6. "Basic Research in the Navy," Report to the Naval Research Advisory Committee by Arthur D. Little, Inc., Vols. I & II, Cambridge, Mass., 1959.

7. Brooks, H., "Past Achievements and Future Foci of the Federal Government in Science," Address to Vicennial Convocation, Office of Naval Research, May, 1966, Official Proceedings.

8. Reiss, H. and Balderston, J., "The Usefulness of Scientists," *International Science and Technology*, No. 53 (May, 1966), pp. 38–44; "Motivating Scientists," *International Science and Technology* (June, 1966), pp. 93–101.

9. Bass, L. W. and Old, B. S., eds., *Formulation of Research Policies*, AAAS Publication No. 87, Washington, D. C., 1967.

10. Sutherland, G., "Some Aspects of the U.S.A. Today: Science," *American Scientist* 55:297–310 (September, 1967).

11. Handler, P., "Academic Science and the Federal Government," *Science* (September 8, 1967), pp. 1140–46.

12. DuBridge, L. A., "University Research," *Science* (August 11, 1967), pp. 648–50.

13. Comments on Department of Defense Project Hindsight, *Science* (November 18, 1966), pp. 872–73; (December 2, 1966), p. 1123; (June 23, 1967), pp. 1571–77.

14. *AFOSR Research* (AFOSR, Arlington, Va., 1967) AFOSR 67-0300, Clearinghouse for Federal Scientific and Technical Information, U.S. Department of Commerce, Springfield, Virginia 22151.

15. Tanenbaum, M. and Committee Members, "Report of the Ad Hoc Committee on Principles of Research-Engineering Interaction," National Academy of Sciences–National Research Council, Materials Advisory Board, Publication MAB-222-M, July, 1966; see also Tanenbaum, M., "Relevance and Responsibility," pp. 68–78 of this volume.

16. Morton, J. A., "A Model of the Innovation Process," in *Proceedings of a Conference on Technology Transfer and Innovation* (May 15–17, 1966), NSF 67-5.

17. Piore, E. R., "Science and Technology in Industry," pp. 1–8 of this volume.

18. Price, W. J., "Concerning the Interaction Between Science and Technology," *OAR Research Review* V, No. 10 (December, 1966); also published in *Cryogenic Technology* 3, No. 4 (July–August, 1967), pp. 141–43.
19. Price, D., "Is Technology Historically Independent of Science? A Study in Statistical Historiography," *Technology and Culture* 6, No. 4 (Fall, 1965), pp. 553–68.
20. Sherwin, C. W. and Isenson, R. S., "Project Hindsight," *Science* 156, No. 3782 (June 23, 1967), pp. 1571–77; also "First Interim Report on Project Hindsight," Summary, June 30, 1966, revised October 13, 1966. AD 642-400, Clearinghouse for Federal Scientific and Technical Information, Springfield, Virginia.
21. Myers, S., *Technology Transfer and Industrial Innovation*. NSF Contract C321, February, 1967.
22. Brooks, H., "Applied Research, Definitions, Concept, Themes," *Applied Science and Technological Progress*. A Report to the Committee on Science and Astronautics, U.S. House of Representatives by National Academy of Sciences, June, 1967, Superintendent of Documents, GPO, pp. 21–55.
23. Rosenbloom, R. S. and Wolek, F. W., "Technology, Information and Organization: Information Transfer in Industrial R & D," Harvard University, Graduate School of Business Administration, June, 1967.
24. Allen, T. J. and Cohen, S. I., "Information Flow in an R & D Laboratory," National Science Foundation, NSF 217-66, August, 1966.
25. Marquis, D. G. and Allen, T. J., "Communications Patterns in Applied Technology," *American Psychologist* 21 (1966), p. 1052.
26. Letters to the Editor on DoD Project Hindsight, *Science* (January 27, 1967), pp. 397–98; (September 29, 1967), p. 1512.
27. Allison, D., "The Growth of Ideas," *International Science and Technology* (July, 1967), pp. 24–32; (September, 1967), p. 1512.
28. Goldman, J. E. "Role of Science in Innovation," in *Proceedings of a Conference on Technology Transfer and Innovation* (May 15–17, 1966), pp. 21–30, National Science Foundation, NSF 67-5.
29. Price, W. J., "Planning Phenomena-Oriented Research in AFOSR," in *Planning Phenomena-Oriented Research in a*

Mission-Oriented Organization, 12th Institute on Research Administration, The American University Center for Technology and Administration (April 24–27, 1967), Washington, D. C.

30. Brooks, H., "Science and the Allocation of Resources," *American Psychologist* 22:3 (March, 1967). Paper delivered at 1966 Annual Meeting of American Political Science Association New York, September, 1966. Copyright 1966 by the Association.

31. Weinberg, A. M., "Science, Choice and Human Values," *Bulletin of the Atomic Scientists* (April, 1966).

32. *Fundamental Problems of Future Aerospace Structures,* summary of a conference April 17–18, 1967, Philadelphia, Pennsylvania, edited by Melvin V. Zisfein. The Franklin Institute Research Laboratories, Philadelphia, 1967. AFOSR Report 67-2175.

33. AFOSR Coupling Activities, 1966; AFOSR Coupling Activities, 1965; Summary of AFOSR Coupling Activities, May, 1965.

34. Teller, E., "The Evolution and Prospects for Applied Physical Science in the United States," *Applied Science and Technological Progress,* pp. 365–397.

35. Alpert, D., "Applied Science and Engineering in the University," pp. 79–94 of this volume.

5   MORRIS TANENBAUM

Relevance
and
Responsibility

The interaction of science and technology is a hallmark of our modern world. Technology, man's use of natural phenomena to serve his ends, predates recorded history. The study of science and the development of the scientific method are more recent but also ancient. That science and technology are interrelated has long been evident. Indeed, along with man's basic curiosity, one of the major stimuli to the study of science has been the effort to understand the natural phenomena on which technology is based. In addition, research relies on technology for its tools. Technology, on the other hand, builds on science but also often reaches beyond scientific understanding, in controlling and using natural phenomena which science cannot explain.

However, over the centuries when interactions between science and technology did occur, they were generally tangential and more of intellectual interest than utility. It is only in the last several decades that the interaction between science and technology has become direct and crucial. Simi-

Morris Tanenbaum: Director of Research and Development at Western Electric Company's Engineering Center, Princeton, New Jersey.

larly it is only in the last few decades that we have concerned ourselves explicitly about how these interactions can be stimulated.

Significance Breeds Concern

The source of this concern is clearly related to the very large role that both science and technology have assumed in our everyday lives. To an ever increasing degree science, in its effort to explain the universe, and technology, in its effort to control and change the universe, have become a natural part of daily life. This is a direct result of the successes which science and technology have had in achieving their objectives of knowledge and service.

As a consequence, the stakes in improving the interaction between science and technology have become very high. Past experience has amply demonstrated the impressive advances that occur when technology builds upon the forefront of scientific understanding. This increases the incentives for improving the ability to interrelate. These past successes and our increasing appetite for the fruits of science and technology, have caused us to increase our investment in these activities. This, in turn, brings growing pressures to insure that the higher level of investment does indeed yield the anticipated fruit.

For these reasons, interest in the interactions between science and technology and how they occur has become intense. Furthermore, the subject itself excites interest. There is an aura of mystery about why and how the interactions take place. Science is widely cultivated in a variety of institutions. Yet some are more successful than others in stimulating technology. Similarly, scientific research may open new vistas of understanding *without* providing fertile ground for the growth of new technology. In total, the patterns by which positive interactions between science and technology take place have been elusive and obscure.

These problems have been of special interest to those who are responsible for determining the funding of science and technology. Quite naturally, this interest has also been shared by scientists and technologists themselves. This mutual concern has stimulated a number of studies in an effort to understand how science and technology interact. The studies have included thoughtful inquiries by Congress and by the executive branches of the federal government. The aid of national institutions and professional societies has been enlisted. These activities have illuminated various parts of the problem and have provided much additional data and some additional understanding. However, as the chosen topic of this meeting suggests, there is still much to learn.

A Case History Approach

Among the various approaches to the general problem has been the investigation of specific cases where fruitful interactions between science and technology have occurred. Some of these case studies have attempted to define not only the specific events which occurred but how and why they occurred as they did. These latter questions are especially difficult. Nevertheless, the attempts to answer them can provide insight into the nature of the problem. The results of a study of the Materials Advisory Board of the National Research Council, National Academy of Sciences, will permit me to illustrate this.[1]

This specific study was performed at the request and with the support of the Advanced Research Project Agency (ARPA) of the Department of Defense. Its object was to examine several instances in the area of materials where an interaction between scientific research and engineering resulted in new technology of important commercial or government use. In addition to documenting the case histories, ARPA requested that the cases be examined in an effort to illuminate how interactions between research and engineering occur.

The Materials Advisory Board approached this assignment by enlisting the aid of a group of scientists and engineers with experience in the support and performance of scientific research and engineering. Each participant selected a subject where he believed that interactions between scientific research and engineering had led to a significant technological achievement. He then attempted to document in detail what had occurred from the initial generation of the pertinent scientific knowledge through to the final technological application based upon that knowledge.

In this way, ten case histories were generated. They concerned developments in metals, ceramics, and synthetic polymers involving structural, electronic, and other applications. After generating a chronological history of each development, the authors searched the histories for recurring patterns of events and circumstances and attempted to find characteristic factors which were common to more than one case. They examined characteristics of the principal individuals who contributed to the material development (e.g. level of formal education, scientific or engineering background, etc.), characteristics of the environment in which the work took place (e.g. type of institution, its organization and communication patterns, sources of funding, etc.) and characteristics of the problem itself (e.g. magnitude of the technical difficulties, visible need for solution, etc.). They then identified those factors which occurred in the cases with sufficient frequency to merit their citation and more thorough scrutiny.

The Motivated Individual

The factor which impressed the investigators most strongly was that in nine of the ten cases studied, *the explicit recognition of an important need* was identified as a major and recurring factor in bringing about the research-engineering interactions. It was very rare that basic research by itself produced a technological opportunity which was quickly recognized and developed. By far the most dominant mode was the situation where an urgent need stimulated a search through prior basic knowledge to find a solution. Perhaps more significant in these cases, it was an individual with a well-defined technological need who initiated the research-engineering interactions, beginning the dialogue that led to the positive interaction between science and technology.

It was similarly interesting that in most of the cases the science that led to the technological solution was available before the dialogue began. It was rare that the technological need directly stimulated the generation of science which was used to solve the technological problem.

In the majority of the cases, the fruitful interactions which took place occurred between organizationally independent groups which were also frequently geographically separated. Again it was usually the individual with a well-defined need who bridged these gaps and maintained the continuity of communication.

The investigators explicitly searched for the presence of a number of other factors. They determined that *flexibility of support was a significant factor in many cases,* since major changes in plans and direction were frequently required before a final technological solution was achieved. They searched for detailed paths and patterns of interaction and were impressed with their inability to describe a common "road map." Indeed the report states that a principal characteristic of the study was ". . . the large heterogeneity of the environments, personalities and problems . . ." and that ". . . research-engineering interactions take place within varying structures of organization and involve complex human events. . . ."

As a member of the study group, I left this experience with increased humility concerning our ability to find unique methods to improve interactions between scientists and technologists. However, along with most of the other members of the study panel, I was deeply impressed by the importance of individual motivation in the cases which we explored. The recurring appearance of an individual who was committed to solve a specific problem seems to merit the most studious consideration in any discussion of the interactions of science and technology. It has important implications concerning the role of individuals and institutions in these interactions. This, in turn, has important implications concerning the support of these individuals and institutions.

Individual Relevance and Institutional Responsibilities

Science and technology are each professions which require the highest level of intellectual effort and personal devotion. Indeed, where science and technology are closely interrelated, the techniques and content of the daily effort of the scientist and the engineer may be very similar. Frequently, a brief examination of the individuals at work will not distinguish between them. However, a discussion of their objectives will generally illuminate the difference. The values by which each makes choices when he reaches a branching point in the train of his investigation are substantially different. These values determine the end point of his studies and they are crucial to his interactions with other professionals and with his environment at large. They determine for each individual that which is *relevant* to his interests. In general he will respond most actively to that which he recognizes as relevant.

The origins of these individual values are complex and important subjects for continuing study. However, there is another aspect of this issue that is more susceptible to definition and influence. Most scientists and engineers are part of organizations and institutions which are themselves characterized by values and objectives closely correlated with the *responsibilities* of the institution. This, in turn, provides much of the environment in which the individual must function and, therefore, can affect individual values. Furthermore, from the viewpoint of the support and stimulation of science and technology, the institution provides a convenient focus for considered action. Thus, the consideration of institutional responsibilities in society may yield useful guides for supporting a proper interaction between science and technology.

Modern society is characterized by a variety of institutions which have achieved responsible roles. One of the most durable of these institutions is the university. There are two principal roles that are generally ascribed to the modern university. One of these is the generation of the broad base of scholarly knowledge which society requires for its development and fulfillment. This includes a knowledge and understanding of natural phenomena which form the content of science and the foundations of technology.

In addition, the university through its educational function is responsible for transferring this knowledge to the young men and women who will form and shape tomorrow's society. In this role, the university is the principal agent for supplying society's most essential components, its educated people.

Industry and government are among the other major modern institutions. Industry traditionally, and government more recently, have developed the responsibilities for providing the goods and services which society desires. These goods and services are based increasingly on technology. In fulfilling these roles, industry and government require the knowledge and the educated people generated by the university.

To the extent that these definitions of institutional responsibilities are valid, they lead to general conclusions regarding the relevance of the activities of their individual members. From this point of view, the scientific research carried on in the university may be considered to be relevant that serves to support the educational function and to expand the broad base of knowledge of natural phenomena which underlie man's perception, understanding and control of his environment. Experience has shown that these two activities may well be symbiotic and that the pursuit of research to expand the base knowledge is one of the most powerful vehicles for education. However, although research and education may be symbiotic, this happy relationship is not guaranteed. If conflict should arise, a choice must be made. In this unhappy case, I would hope for a decision in the favor of education, for society depends wholly upon the university for this function.

In industry that effort is relevant which expands the goods and services that society desires. The basic scientific research carried out in industry is relevant which promises to expand the base of understanding for technology in fields of interest to the industry. The expansion of knowledge for its own sake is not sufficient.

The role of government has grown to overlap both the university and industry since education, defense, health, and other functions are among the services where our society has defined an important responsibility for government. For this reason, the questions of relevance are of special concern. Actions in the support of one function may not serve another. Thus, it is important to separate these objectives clearly and consider actions in terms of the desired balance between these functions.

Patterns of Support

These definitions of relevance within the context of institutional responsibility can lead to general criteria for the support of science and technology in these institutions. This, in turn, will importantly influence the environment for the interaction between scientists and engineers. At the university, support must be directed to those programs which are consistent with the stimulation of high quality education and which promise to advance the broad bases of man's knowledge. Industry, on the other hand, must support that science which has the most promise of expanding the base of knowledge in areas relevant to the industry's technology.

Of course, the proper use of such criteria requires great thought and judgment since the proper definition of relevance requires insight and foresight. The selection of suitable research theses deserves the judgment of the most capable educators. Similarly the selection of broad areas of science which will provide the basis for future technology requires the judgment and imagination of the most capable industrial scientists and administrators. Government must face both problems and make choices based on the balance of its objectives.

These decisions also require extensive interaction between institutions. For example, the university must interact with industry and government in order to evaluate the trends of educational needs and to judge the success of prior educational efforts. Industry and government must interact with the university in order to determine the progress of scientific understanding and evaluate the relevance of newly developing areas.

In this light, society in general and industry and government in particular, must continuously reexamine the bases by which support is determined for both scientific and technological interests and especially for the critical functions of the university. There is evidence that in the past support of academic research has often been justified on the basis of its relevance to technological needs. Fortunately, through the efforts of thoughtful individuals among the agents of support and the recipients, areas were discovered and programs defined which could be properly justified on the basis of technological relevance and remain consistent with the university's fundamental responsibilities of education and the generation of basic knowledge. However, I question whether this approach will be adequate for the future. Many of the strains which currently exist and which lead to current reappraisals of the "success" of academic research may well result from a schizophrenia in policies of support.

Similarly, industry must reexamine its methods of support and interaction with the university. Industry has a most direct and immediate interest in the university's educational processes. This interest has grown vigorously as the demand for industrial scientists and engineers has increased. At the same time, industry's direct support of the university has decreased in relative magnitude although its overall support remains high through its contributions to tax revenues which find their way to the university. Industry must reexamine its methods of direct support in order to assure that these funds are used most imaginatively and effectively to stimulate the educational processes which serve it. In addition, as responsible citizens, industry must contribute to the formation of the most effective government policies in this area.

Admittedly, these institutional considerations do not directly attack the problems of individual motivation which seem so critical to interactions between science and technology. Nevertheless, they do contribute importantly to the individual's environment and this will affect his values and judgment. Of special importance, a national recognition of the differing institutional roles will assure the development of consistent patterns by which we support our essential institutions and their individual members. In our society, this is perhaps the most acceptable way to influence individual motivation on a general scale.

Looking Forward

Within our lifetimes the interaction of science and technology has grown to be of critical significance to the university, to industry, to government, and to every member of our modern, technologically based society. It behooves us to study well this area where fundamental objectives, deriving from a variety of differing individual and institutional values, must be harmonized to achieve the maximum benefit of human abilities. If we can understand these differences in objectives and their similarities, if we can define a societal context where these objectives can be strengthened in their fundamental directions and yet encouraged to interact harmoniously, we will have established patterns which will insure the most effective and appropriate roles of science and technology in the future.

REFERENCE

1. Tanenbaum, M. and Committee Members, "Report of the Ad Hoc Committee on Principles of Research-Engineering Interaction," National Academy of Sciences–National Research Council, Materials Advisory Board, Publication MAB-222-M, July, 1966.

6 DANIEL ALPERT

Applied Science and Engineering in the University

During World War II, a handful of university scientists assumed a self-imposed responsibility which was to change man's notion of the role of science and engineering in society. One aspect of this remarkable event was that a number of university scientists stopped practicing their profession, stepped out of their usual role, and became what would be called now systems engineers, component developers, and systems managers. Another aspect of the event is that a group of laboratories was formed—the radiation laboratories at M.I.T. and at Berkeley, the national AEC laboratories at Oak Ridge and Los Alamos, etc. In short, a new form of social organization, the independent, government-sponsored, not-for-profit laboratory was invented, an institution which was to continue to make a major impact on our society long after the scientists went back to their classrooms and laboratories. Out of these wartime efforts came radar, the proximity fuse, the atomic bomb, and the nuclear reactor.

Daniel Alpert: Dean, The Graduate College, University of Illinois, Urbana, Illinois.

As significant as these developments seem in retrospect, the level of effort and dollar volume involved only a small fraction of the total technological commitment during the war. The vast array of other weapons, tanks, artillery, ships, and aircraft was designed under more traditional forms of engineering management within the industrial complex of the nation. But the spectacular developments of radar and the atomic bomb were in the postwar years to make a profound effect on the way of life of the scientific and the engineering community, not only in the university but in industry as well.

I am sure that much of the discussion at this symposium will be reflections upon the chain of events which has followed, in large part stimulated by the prestige gained during the war by these scientists turned engineers. Given what seemed in 1946 a blank check to support their research, the majority of university people and many in industry returned to their original culture, basic research in science. As a result of continuing support for the past two decades, the basic sciences have prospered to the extent that the United States is the acknowledged world leader in almost every field of the natural and life sciences. As a by-product, the prestige of the scientist in academia has risen to such a high place on the pecking order that virtually all other disciplines or departments—from engineering and medicine to the social studies and even humanities—have taken on the science culture to a greater or lesser degree. In engineering it is often hard to say whether the fundamental sciences have been incorporated or whether they have replaced engineering in the educational program. As Harvey Brooks has pointed out, "After decades of inadequate attention to basic science and inadequate recognition for scientists, the United States may have overreacted. Raising the status of applied science in universities has become a real problem. . . ."

Inevitably, at this point in the discussion, there will be

questions about what I mean in my distinction between basic and applied science, or between science and engineering. I would prefer not to go into a long semantic discussion; almost everyone who addresses this topic seems impelled to spell out on his own the difference between basic or phenomena-oriented research on the one hand and applied science and engineering on the other. However fuzzy the interface may be, I am convinced that some such classification is needed to protect many basic research efforts from unrealizable expectations on the part of those providing financial support. If nothing else, we owe it to our sponsors (i.e., society) to make a reasonable estimate of whether our research effort is likely to pay off in terms of a solution to a human problem in the course of the next year, the next decade, or the next century. In my thinking, I have found it useful to differentiate by noting that the scientist selects his problem from among the puzzles posed by nature. Though the applied scientist or engineer may also be called upon to expand the understanding of natural phenomena in the course of his work, his problem was posed in the first instance not by nature but by society. The engineer may often start with a solution in search of a problem, rather than vice versa; but he is, nevertheless, seeking to apply his findings to the real world of technological utilization. I will not dwell here upon the distinction between engineering and applied science; it seems to me that when the engineer deals with state-of-the-art technology he is performing an engineering function; when he is called upon to expand the state-of-the-art to provide a solution, he must usually do some applied science.

I have mentioned the trend to elevate science and basic research to the top rung on the prestige ladder of the university. As valuable as the trend has been to introduce basic science and mathematics into the engineering curriculum, I feel there are serious dangers, not only to engineering but to science as well, when the culture of science replaces the culture of engineering in the engineering colleges of our universities. I am going to try to elaborate on this observation and to discuss some alternatives which lie before us.

As I have said, the intellectual life of our engineering colleges has benefited greatly from the incorporation of the basic sciences into the curriculum. However, in many cases, the pendulum has swung far beyond this point. High status and rewards have been assigned to the purist who is more inclined to develop mathematical models than to verify whether the models relate to the real world. Appointments and promotions are all too often based on the number of papers published without regard for relevance rather than on engineering accomplishments or inventions. All too often, engineering graduate students are trained in an environment *dominated by a science culture* in which the problems assigned are not those posed by society but those posed by the closed group representing a small subdiscipline. Frequently the engineering department contains a number of insulated professors whose cultural ties lie in science departments elsewhere on campus and not with each other. Note that I emphasize the words culture and environment because it is not the methodology or implementation of a given investigation which differentiates between pure and applied science.

I have an overriding concern that, if our engineering departments assume a science culture, the university as a whole cannot respond to the need for a valid interface with the problems of society. But why should the university provide such an interface? Why not restrict ourselves, as Robert Hutchins would prefer, to the pure science and humanities, and leave it to other institutions to solve the problems of society? There are several valid reasons for disagreeing with Mr. Hutchins. First of all, as Edward Teller has pointed out, our society needs well-trained applied scientists and engineers, oriented to the solution of problems posed by society. All too many of our nation's applied efforts are now being guided by members of an older generation—my generation—and this leadership is not being replenished by younger people who are willing to tackle difficult and complex problems dished up the way society poses them, as well as puzzles dreamt up by clever professors. We must educate new leaders. And we must recognize that most young people develop a cultural view of their professional careers during their formative years in the university.

Some of my friends hold the view that it is eminently reasonable for the university to train its future engineers in the basic disciplines, then turn over to industry the responsibility for teaching them to solve real problems. After all, hasn't American industry done eminently well in recent years? In many areas, the answer to this question is affirmative. Generally speaking, our high-technology industries, e.g., those in military-space, electronics, and communications, have effectively utilized and completed the education of our engineering graduates in a way which has qualified them for careers in these fields. But in many of our civilian industries, companies have either failed to lure our science-oriented engineering graduates, or, when they have done so, they have failed to provide adequate education or career incentives. In some industries, such as railway and marine transport and shipbuilding, this country has fallen far behind. In others, such as electrical power utilities, the development of radically new engineering solutions has come in large measure from independent, government-sponsored, not-for-profit laboratories. Thus, it is not at all self-evident that our society should turn over to industry the responsibility for all of our nation's engineering—let alone the professional training of our engineers.

Engineering has been defined as the art or science of converting natural resources to the utilization of man at a price he is willing to pay. To the narrow economic definition of "price to the customer" we must now add the dimensions of "price to society"; i.e., the dimensions of social values and of political constraints. As Richard Soderberg has pointed out, "It is worth reiterating that our technological and business system as such is indifferent to the many moral and ethical issues involving difficult collective or social choices which are more than the least common denominator of private value systems. Neither technology nor science will by themselves provide anything but means toward ends. The ends themselves must come from other motivations. To an increasing degree, private interests and benefits come into conflict with collective values which cannot be measured in the marketplace. . . ." It is in this context of the changing values and responsibilities of the technological enterprise that I question the wisdom of turning over to others the professional education of future leaders in applied science and engineering.

It is my opinion that the involvement of our professional schools in the solution of real problems is good for the university. The reasons are both philosophical and pragmatic; they are inherent in the traditions of American higher education, particularly those of the land-grant universities. Two basic assumptions of the land-grant movement were (1) that we should provide a first-rate higher education to a large and increasing fraction of our population, and (2) that our universities should assume an important role in the public service. Some of the unique qualities of American universities at their best are attributable to these society-oriented objectives. I would also argue for a strong interface between the university and society on purely pragmatic grounds. It was the commitment to public service and the solution of human problems, largely in agriculture, which helped to retain the grass-root support of the people of this state for their land-grant institution during its first hundred years and made it possible to build a major university out here on the Illinois prairie. The need for support and understanding at the national level is also essential. If we are to continue to receive massive federal support for graduate education and research, it is essential that we develop a better interface between our mission-oriented society and the discipline-oriented university.

If you accept my premises that the university can and should relate to the problems of the real world, what problems are faced by the professor of applied science or engineering? How does the scientist relate to a modern technological or engineering effort? In addressing these questions, I think it is useful to start by trying to characterize the structure and ingredients of a modern, mission-oriented, technological effort. I attempt this as follows:

1. A modern applied effort involves a team with a highly qualified technical leadership. Since the team usually involves specialists in a number of widely different fields, it goes without saying that the technological project calls for management. Frequently the technical leaders are the managers. Having said that an applied effort calls for an interdisciplinary team, I feel called upon to say what it is not: it is not an interdisciplinary committee, nor is it a group of people from different disciplines working in the same building or paid from the same budget.

2. The applied effort is expensive; it often requires sophisticated and costly services and facilities; it calls for decision-making flexibility at the local level in the expenditure of funds.

3. The applied research project is directed toward specific goals. It calls for the formulation of a strategy specifying both technological objectives and valid approaches to a problem which is more often stated implicitly than explicitly.

4. It involves a commitment to a time scale which is known by and acceptable to the sponsor, whether the sponsoring agency be the parent company, a government agency, or a laboratory management.

5. In addition to a highly motivated leadership, the applied effort calls for a group of professional personnel who have expert state-of-the-art proficiency in the various fields involved and who provide continuity to the project by a total commitment to its goals. (This can seldom be provided either by professors or graduate students.)

Now, before addressing myself to whether an applied effort can be carried out within the university, let me try to enumerate some of the various ways in which a professor may relate to such an activity without regard to where it is located. I believe that much of the confusion about technological spin-offs from the university or about how universities should relate to industry is generated by a very fuzzy notion in some quarters as to how scientists anywhere—even when working for the same corporation—relate to a mission-oriented project.

I think it is worth trying, even with a broad-brush treatment made necessary by brevity, to explain this complex process, if only to challenge or to limit fruitless efforts at communication. All too often such efforts are generated by well-meaning administrative officials or public relations men who have never taken part in a valid dialogue with either a university scientist or an industrial applied scientist, but who are promoting endless conferences and useless paper shuffling, ostensibly to generate dialogue between them. For example, if we assume as a model for such a dialogue the relationship between the Chamber of Commerce and community businessmen, the efforts are destined to generate noise without signal and expectations which can result only in disillusionment.

I have tried to characterize four ways in which a valid professional interaction can take place between an individual professor and a modern, technological, mission-oriented effort:

1. *Scientific consultant.* In this case, the mission-oriented team must be sophisticated enough to formulate questions for the specialist in such terms and at such a level of understanding that he can respond as a specialist in his own code language. Obviously, in this case, the burden of real understanding of how science fits in is carried by the applied team; it goes without saying that if they are good enough to pose the question they are also good enough to know the people in the discipline, and to evaluate the answers.

2. *Engineering consultant.* In this case, the individual professor undertakes not only to answer questions but also to pose them. He must be sufficiently knowledgeable to be able to formulate the proper questions on the basis of his understanding of and commitment to the solution of the overall problem. This is how I differentiate the engineering consultant from the scientific consultant. The burden here is on the individual and calls for much more than expertise in a narrow field of specialization. It calls for the type of professor whom Gordon Brown must have had in mind when he said, "An engineer is a scientist who has completed his education."

3. A third kind of relationship to the real world is that characterized by the activities of many professors of architecture in the real world of building design and construction. They maintain their own professional practice; in the technological context we would call them systems designers or systems engineers. This kind of a relationship brings the professor directly into the real world. However, with the growing complexity of modern technology, systems engineer-

ing and even architecture itself have become so sophisticated as frequently to call for a team effort beyond the capability of a single professor or his students.

4. The fourth style of relationship between professor and mission-oriented effort is that in which he takes a leadership role in the project. Though examples are not numerous, this *has* happened in universities and in some cases such projects have been unusually successful. One of the very first computers built anywhere, ILLIAC I, was engineered and constructed here on the University of Illinois campus. We now have teams working on ILLIAC III and ILLIAC IV. There are also various mechanisms whereby a professor may play a leadership role in an applied science project located away from the university campus, in either an industrial or not-for-profit laboratory. I believe that George Pake and William Linvill will describe such efforts carried out in distinctively different ways.

I am sure that my listing of four possible styles for interaction between a professor and an applied science effort does not cover all of the possible modes; for example, I did not list technical management consulting; furthermore, there may be mixtures, involving more than one pattern. But I think it is of value to consider the validity and effectiveness of these particular modes of interaction and to describe some of the problems involved.

At the outset, it is my general observation that communication in all of these categories is typically very personal and not effected *via* journal articles or voluminous reports except as an aid to the communication process. Technological transfer is seldom, if ever, transmitted through the mails. It is carried out in a variety of formal and informal human contacts, often by word-of-mouth, and on the basis of very casual interactions. It is probably for this reason that many of the engineers interviewed in the course of Project Hindsight could not even identify the original inventor of an important component incorporated in their systems or specify the original scientific paper in which an important idea was contained.

In considering the specific styles of interaction, let me start with the scientific consultant. From what I have said, it should be obvious that it is relatively straightforward for a science-oriented professor to relate in a consulting role to a highly sophisticated technological effort, providing the latter group has a need for understanding in his field. This is one reason many of our top-flight science professors find it both educationally and financially profitable to maintain a consulting relationship with a modern industrial laboratory. I think it is also evident why it is difficult, if not impossible, for a scientific consultant to relate to a technologically backward company. Almost by definition, such a company is not in a position to state its problems in scientific terms. Thus, it is far easier for a scientist to relate to a sophisticated technological effort located a thousand miles away than to relate to a technologically obsolescent effort across the street. In the professional interaction between a scientist and a technological project, the cultural impedance match is far more significant than the geographical barriers.

Not only is it difficult for a technologically obsolescent company to state its problems in scientific terms; it is often limited in ability to make a valid statement of its problems even in technological terms. Thus, unless the professor-consultant undertakes not only to state but also to solve the technological problems of an obsolescent industry, he too may find it difficult to bridge the cultural gap. To do so, he may be faced with the necessity of setting up a mission-oriented project of sizable proportions in order to achieve success. The most likely consequence to the professor who tries, *with limited resources*, to relate to an obsolescent industry is that he will become obsolete himself. And however highly such a professor may be valued by the company because he tries to cater to it on its own terms, he will be low man on the totem pole in the academic community not only because he is doing something useful but also because he has lost contact with the frontiers of his own profession.

It is for these reasons of cultural mismatch that the State Technical Services Act has not found it easy to develop widespread interaction between the university and industries which are in need of more sophisticated technology. As I have suggested above, it may call for a mission-oriented team effort of major proportions within the university in order to break the cultural barrier. An outstanding success story in this area is the interaction between the Civil Engineering Systems Laboratory of the University of Illinois and the building design and construction industry. In this case, some of the key professors in the Civil Engineering Department had built a highly skilled and broadly knowledgeable interdisciplinary team which could provide for the industry a powerful approach to its problems, using modern time-sharing computer techniques. Equally important, a sufficient number of far-sighted firms in that industry were receptive to exciting new ideas when they were exposed to them. I am convinced that the results will be of outstanding value, both to the industry and to the university. It is important to note that the professional interaction between the leaders of this project with the other members of the very sophisticated computer culture at this university is one of mutual respect. In other words, a truly successful interaction with the outside world can be accompanied by a cultural impedance match between the discipline-oriented specialists and the mission-oriented engineers on the campus. One could turn this around and say it is not a valid interaction with industry unless the university-applied effort also is culturally tied in with the university. The group team effort that is involved in this civil engineering development is at the forefront of the engineering profession, and the graduate students who relate to it are getting a first-rate education.

I have suggested that some strong interactions between engineering professors and industry are culturally compatible within the university and some are not. At the other end of the spectrum we have engineering professors who have chosen to adopt a "science culture" in engineering. If incorporated within reasonable limits, these individuals may provide the necessary impedance match or coupling between science and engineering in the university. However, if engineering departments become dominated by a preoccupation with mathematical techniques or pure science, it is rather doubtful that they will do either good engineering or good science; in any case, it is certain that their students will leave the university with a distorted view of their profession. Indeed, some such students may find that the only other engineering culture of comparable purity—some have called it sterility—is to be found only in another educational institution. If this becomes too prevalent, and there is a marked similarity here with other professional schools (medicine, education, business, etc.), there will be little meaningful interaction between the university and society. In all of our professional schools we must try to choose a path which avoids the obsolescence associated with a nose to the grindstone approach to practical problems and the sterility associated with avoiding them altogether.

In my concluding remarks, I would like to address myself to the problems of the administration of mission-oriented research in the university, especially to react to Alvin Weinberg's thesis that, since society is inherently mission oriented and the university inherently discipline oriented, one should view applied research as to be done somewhere else, in a separate organizational framework such as the national laboratories. I have acknowledged from the outset that the independent, not-for-profit laboratory is one of the important social inventions of our times. Some of the very significant technological and social innovations since World War II have originated in these institutions. I also acknowledge that it is remarkably difficult to administer viable mission-oriented projects in a manner compatible with the science-dominated and discipline-oriented departmental structure of the university. But I am not ready to acknowledge that there should be an institutional and geographic separation of these activities. Indeed, if the professor is to be a citizen, it seems to me that he will have to relate to society in many ways.

Three of the most exciting areas of research activity on this campus illustrate successful examples of mission-oriented engineering projects with University of Illinois professors in leadership positions. I have already cited the program in structural design and management in our Civil Engineering Systems Laboratory; a second is the ILLIAC IV project, aimed at the design and construction of a uniquely powerful parallel computer; the third is the PLATO program in computer-based education that originated here in the Coordinated Science Laboratory. There are numerous others.

It has not been easy, within the rigid department-oriented structure of the university, to incorporate such large and expensive mission-oriented projects, and I do not pretend that all of the problems have been solved. My own efforts at this university were for eight years devoted to establishing at the Coordinated Science Laboratory an administrative impedance match with both science and engineering departments. The effort could not have been possible without a block grant from the three military services which gave the director the prerogative to initiate new applied efforts without costly and ineffective efforts to justify them at their earliest incubative period, and to support them with some significant professional effort.

If we are to accept funds from mission-oriented agencies, ostensibly to solve the increasingly complex problems which they face, some involvement with major interdisciplinary projects may represent the only valid response. And this is probably the case in many fields, including the recently recognized urban problems. Any valid applied effort should satisfy the demanding criterion that it present a cultural match with the best in the university community; that is, it should stimulate a supportive interaction between scientists and engineers on the same campus. To accomplish this, and simultaneously to justify the support from the sponsors, we may in the future have to encourage more block grants, supporting both pure and applied science.

It will be a continuing challenge to provide an administrative as well as a cultural impedance match. To overcome the internal difficulties of fitting mission-oriented functions within the traditional forms of the university will certainly call for a continuing supply of uniquely talented scientist-engineers with a flair for administration and to elevate administration to a respectable role in the university. We will have to do considerable soul-searching before we have faced up to these challenges.

If it is true that the university finds it difficult to adapt its form to the new functions demanded by a modern technological society, it is also true that there are difficulties facing the free-standing, not-for-profit laboratory, especially in maintaining viability after its initial goals have been reached. Although the independent "national laboratory" has a more viable administrative structure within which to set up major mission-oriented projects, there are numerous examples of good scientists from the not-for-profits who have chosen to return to academia to renew inspiration or to tackle new problems. This flow between the universities and the national laboratories is to be encouraged, and it should be a two-way flow. It might be even more effective if the two kinds of institutions were geographically adjacent. Weinberg recognizes the validity and desirability of the day-to-day contact between his mission-oriented laboratory and the University of Tennessee. But since it was difficult to move the Oak Ridge Laboratory to Knoxville, portions of the University of Tennessee have been moved to Oak Ridge. Obviously, the geographic barrier which isolates many of the major national laboratories makes it difficult to maintain supportive day-to-day interactions with the university. I leave in the form of a question whether it is feasible or desirable to have a much closer relationship, geographic as well as cultural, between universities and independent, not-for-profit institutes.

In any case, it seems to me that the university has a responsibility for graduate education and research in applied science and engineering. Perhaps we should seriously consider a new type of relationship between universities and not-for-profit laboratories which would make for a compatible administrative structure for both and would broaden the framework for interaction between the intellectual community and society.

7 GEORGE E. PAKE

A University-Industry
Joint Venture
in Applied Research

Some years ago the leadership of the Advanced Research Projects Agency of the Department of Defense decided to attempt an experiment to increase the interaction between science and technology. Perhaps because both the Director and Deputy Director of ARPA were physicists, the pilot program was called the "coupling" program, with the object being to seek out and test new methods of tightening the coupling between science and technology or, as we sometimes express it, the coupling between basic science and applied science.

Clearly another kind of coupling was also visualized—that between different kinds of institutions. Most of the nation's basic research occurs in universities; most of the technological applications occur in the industrial world. Are there mechanisms by which association of these various kinds of institutions can enhance the interaction between science and technology?

To answer these questions and to uncover and test

George E. Pake: Executive Vice-Chancellor and Provost, Professor of Physics, Washington University, St. Louis, Missouri.

possible mechanisms, ARPA requested preliminary descriptions of coupling arrangements and programs from teams of two or more institutions. The areas of science and technology suggested for emphasis by ARPA related to a number of kinds of useful materials for which we seek optimum properties with respect to particular applications. Selection of the precise area of concern and of the coupling mechanism was left to each team as it drew up its preliminary plan. Some 80 teams, I believe, submitted these preliminary plans, and about a dozen or 15 were asked to elaborate their plans into definite proposals, from which four or five would be selected for award of contract support.

One of the teams selected was a Monsanto Company/Washington University team. This team has now completed a little more than two years of activity (since July 1, 1965). I would like to describe the origin and nature of our coalescence into a team, to indicate the major problems we faced, to indicate how some of them have been solved, and then perhaps to make a few tentative extrapolations in an effort to estimate whether our particular pilot model is a success. The work proceeds under a contract of one million dollars per year, monitored by ONR.

Background Relationship of Washington University and Monsanto

The metropolitan St. Louis region has an intellectual, cultural, and even a growing economic unity. (Accidents of political history, it must be admitted, created a hodgepodge of *governmental* jurisdictions: three counties in Illinois, three counties and the city of St. Louis (itself *not* a county) in Missouri, and something near 200 incorporated communities.) In this otherwise integral metropolitan area, there has been a continuing relationship between Washington University and the major industries of St. Louis, including Monsanto. Some of the relationship is obvious: Washington University graduates take employment with Monsanto, Monsanto employees sometimes send their offspring to Washington University, some Washington University professors have been individual consultants for Monsanto, and some Monsanto executives have served from time to time on the University's Board of Trustees. All of the interactions are made relatively easy by the fact that both Washington University and the Monsanto campus are in suburban residential areas on the west side of the metropolitan area. It is a relatively pleasant 15-minute drive by automobile between the two campuses.

Another kind of association has seen the two institutions in a developing concern for the economic and technological growth of the region. Monsanto professionals and executives, on one hand, and Washington University faculty and administration, on the other, have been in continuing discussions with each other about the problems of community development. The ARPA announcement of the "coupling" program arrived where informal association and communication was already strong; it was not necessary to scurry around and crank up an ad hoc relationship.

It *was* necessary to distill from discussions of past ideas and from current capabilities a specific and practical plan. The experiment would have the best chance of success if an area of technical activity were selected which was of genuine mutual interest. The field of high performance composite materials was chosen, and the limitations of the university were perhaps the determining factor here. One of our best engineering departments, Chemical Engineering, had strengths and interests, particularly in polymers, which fitted nicely with Monsanto, and there was already good rapport among the technical people in both institutions.

Let me interject here a word about composite materials. Composites have been known and used by man for many,

many years. Perhaps the most common example is concrete. But the high performance composites in which our joint research team has been interested are typically composed of metal or semimetal whiskers or fibers imbedded in a plastic matrix. For example, filaments of boron, carbon, or even glass have higher stiffness-to-weight and strength-to-weight properties than steel or aluminum. An attempt is made to capitalize on these superior properties of filaments by binding them together with a plastic—often an epoxy type material. It is not surprising that the properties of these composites depend upon such variables as the shape and orientation of the filaments, the volume fractions of the constituents, and of course the filament material and matrix material selected. It is also evident that the physics and chemistry of the interface between the filament and matrix material are fundamental to an understanding of the stiffness and strength of the composite.

Organization

Of course the joint research and training venture that constitutes the Monsanto/Washington University effort on composite materials could not exist effectively without the full concurrence of the administrative leadership of the university and the corporate management of Monsanto. But the real working elements of this association are the Central Research Department of Monsanto and the School of Engineering and Applied Science at Washington University. The program manager is Dr. Calfee of Monsanto, who works closely with Professor Stephen Tsai, Director of the Materials Research Laboratory at Washington University. Both of these men maintain close liaison with Dr. Richard Gordon, Director of Monsanto Central Research, and Dean James McKelvey of Engineering.

The research is divided into three disciplinary areas:

mechanics (stress analysis, fracture mechanics, and testing); physics and chemistry (strong emphasis on interfacial problems); and fabrication and processing. The responsibility for planning and research strategy in each of these areas is in the hands of two-man committees, with each committee having a member from the university and a member from the company. These six individuals constitute the Research Council of the Association. The Council meets monthly or oftener, under the chairmanship of Dr. Calfee, and functions as a policy planning and advisory group.

The Materials Research Laboratory at Washington University

In August of 1966, and in large part because of the M/WU ARPA Association, Washington University established the Materials Research Laboratory (MRL) as a division of the School of Engineering and Applied Science. The function of the laboratory is to conduct research in structural materials, in such a way as to complement the academic programs of the School of Engineering and Applied Science. Faculty members and graduate students from any department may participate in the activities of the laboratory which has, however, some professional technical staff of its own.

Because severe space problems at the university will exist until the fall of 1969 when the new engineering building will be completed, the staff of MRL's composites program occupies laboratory and office space at three different locations. It occupies 2500 sq. ft. in the engineering complex on campus; it rents about 1300 sq. ft. near Monsanto, and it occupies a module at the Monsanto Research Center.

Personnel involved include six full-time faculty members from two departments of the Engineering School, plus three affiliate professors, two of whom are, in fact, members of the Monsanto research staff. Nine graduate students are currently involved, eight are candidates for the doctorate and one is a candidate for the masters degree; two masters degrees have been given since the inception of the program.

There are approximately a dozen engineers and scientists of the Monsanto Central Research Department actively committed to the program. Two of these are also affiliate faculty in our Engineering School, as mentioned above.

Problems Encountered in Joining the Two Institutions

At the outset there were administrative discussions to establish the procedures and ground rules in the areas one would expect might be troublesome: a three-cornered patent agreement, the question of possible affiliate faculty appointments for Monsanto staff, the matter of individual consulting arrangements for university faculty, safeguarding company proprietary information, and, of course, the geographic distribution of the research effort between the two (temporarily three) locations. It undoubtedly cannot yet be claimed that all of these matters have been ideally resolved. But, in my own opinion, they have been handled with less difficulty than most of us anticipated.

The patent agreement of course includes the usual cost-free license privileges for the federal government. The more unusual problem is the anticipation of a situation in which both the university and Monsanto have a joint role in a particular patent. This was handled by setting general guidelines for determination of the equities and by the agreement that a joint ad hoc committee from the Monsanto directors and the university trustees would be set up to negotiate the distribution of royalties for any particular patent falling into this category. We have not yet had to test this mechanism, but I think it is highly workable.

We have made affiliate faculty appointments sparingly, with little real difficulty.

Protection of proprietary information for Monsanto appeared to me, at first, to be a serious problem. I did not believe it would be serious in reality, but I was sure that the company would have substantial worries. This is in part already solved within Monsanto, which keeps company-sponsored proprietary research and development separate from its government-contract research which is handled in a wholly owned subsidiary, Monsanto Research Corporation. The ARPA-sponsored program is thus kept separate from the company proprietary sector. If this is done with the Association research on composites, the problem of faculty awareness of proprietary information is really no different from what it has always been under individual consulting arrangements with faculty such as have existed for decades. The company is still free to hire a faculty member as an individual consultant, in accordance with university policy which limits the time a full-time faculty member may devote to consulting. To the extent that this sometimes is done, I should emphasize that (1) ARPA funds do *not* pay for consulting at Monsanto by our faculty, and (2) ARPA contract funds at the university are not and cannot be used to increase the pay rate for a faculty member above the rate stated in his annual letter of university appointment. Students are not involved in proprietary research.

Distribution of the research effort and personnel among the three locations is continually reviewed by the Research Council. It is usually a fairly obvious decision. Special testing or laboratory facilities exist at Monsanto that are not available on the campus, and vice versa. Seminars and symposia are held in both locations. Classes are at the university only.

I might add, in closing this section, that when the Monsanto directors made the "go-no-go" decision on whether to approve the submission of our detailed joint proposal to ARPA, I sat with the Monsanto directors' technical committee both to answer questions and to represent the university's view point. The proposal had been developed jointly by university faculty and Monsanto scientists, and it already met the university's general requirements for a research proposal. The ARPA contract is with Monsanto, and Washington University operates in effect as a kind of subcontractor to Monsanto in accordance with ONR regulations.

Other Activities of the Association

In addition to research and graduate training, our joint proposal set forth a plan to communicate the results of the research. It had two elements:

1. An annual symposium on composite materials research. The third annual symposium will be held on October 26–27, 1967, in St. Louis, and the preregistration promises a large attendance. The second symposium a year ago attracted a national audience of 200 scientists and engineers.

2. After much soul-searching, we have begun as a joint venture, outside the ARPA contract, the Journal of Composite Materials. We found that papers on composites were so widely strewn throughout the technical periodical literature that it was an impediment to progress in the field. The first issue of the journal appeared in January, 1967, and the 1967 volume will total 400 pages. As of June 1, 1967, there were 550 paid subscribers.

Conclusion

I have left until the very last perhaps the most important question. Why should either an industrial corporation or a university go to all the trouble that I have described to set up a joint research enterprise when either could, so it might appear, equally well "go it alone"? Why bother with coordinating committees and splitting up the research over a minimum of two locations?

The answer, we believe, is at the very heart of the topic for this meeting. We are convinced that there should be more rapid and more extensive communication of basic research results *from* the campus and better communication of the nature of applied problems *to* the campus. Our associates at Monsanto are just as desirous of this communication as we are.

Moreover, *engineering graduate education has always had a problem in relating itself adequately to the real world of technology*—for the simple reason that the technological arena is not the university. This is in contrast, for example, to graduate study in physics where, by and large, the universities of the nation are the scene of the major advances in physics. What we seek, therefore, is a way to maintain the essential benefits of the university atmosphere for graduate engineering while at the same time coupling that educational experience more tightly to technology in industry.

It is much too early to draw conclusions about our effort as a general model or even as a particular one-of-a-kind venture. I am, however, highly encouraged, and I thought this an appropriate occasion to tell you a little about it.

8 WILLIAM K. LINVILL

Man,
Technology,
and Society

Introduction

Vast technological changes cause significant shifts in our society and lead to important consequences for the next generation. Man can be either the master or the slave of the new technology, depending upon how he approaches the future.

Technical advances are shrinking the world by greatly expanding man's scope of control and development. As each entity in our society expands its scope of activity, the overlaps become chaotic if they are not foreseen and planned for ahead of time. Continual change and development must be the expected way of life. An unusual flexibility and mobility will be called for from the professional workers in this new day.

Society Must Achieve Coordination While Retaining Free Enterprise

The technological changes ahead are so vast and their consequences are so drastic that chaos is likely to result un-

William K. Linvill: Chairman of the Department of Engineering-Economic Systems, Stanford University, Palo Alto, California.

less some systematic coordination is achieved. Society's problem is to provide systematic coordination while retaining the tremendous advantages of free enterprise. How can this be done?

Comprehensive projections of the future, open to all enterprisers, and an open analysis of possible new joint technological ventures would be a tremendous help in the exploitation of technology to the benefit of man. Broad and continuing education and training are needed to keep men productive in the environment of rapid technological advance. An agency or agencies to provide these functions coupled with the free enterprise system to develop the potentials discovered would probably do the job.

At present no agency or organization serves the future projection, exploration, and broad training functions. What sort of agency might be effective? Such an agency must combine the drive to do and the drive to understand, and it must have the capability for regeneration because it must be effective for several generations.

Universities Appropriately Extended Might Serve the Three Functions

A broadly based university program, properly coupled to action agencies in the world of affairs, might be partially effective in serving the three functions. No university now serves these functions, but with proper help from the world of affairs, they might take on at least part of the job. Further, without the universities' participating in the three functions, they couldn't properly serve their traditional role in the new environment.

What must be added to the present university programs for them to be effective in the future projection, evaluation, and broad training functions? Two vital extensions appear to be necessary:

1. A close, effective coupling must be developed between the university and the world of affairs.

2. A broadly based system program should be developed as a professional discipline.

Coupling between the university and the world of affairs could be achieved through an institute associated with the university but separate from the traditional university program. This institute could carry on projective studies to lay out possible future directions of technological development. It could host joint studies of interdisciplinary teams from government, industry, and academic life to explore new ventures. It could bring fellows from the world of affairs to the university for both projective studies and for exploration of ventures. It could foster programs in which university professors and students would take term responsibility for projects in the world of affairs.

A new professional discipline in system analysis could be added to existing professional school programs. It could develop a training program to combine theoretical work in system analysis with specific practice in the many substantive areas. Comparisons among problems in various substantive areas are indispensable to build this new area. Work in the substantive areas could bring together academicians from the appropriate humanities and science specialties along with practitioners from the world of affairs.

A pilot program along these lines has been established at Stanford University. A few initial projects have been undertaken, and a program of courses and research has been set up. A modest program of internships and industrial fellow projects has been started.

The remainder of this paper will explore these thoughts in some detail and describe our plans more completely.

Exploiting Technological and Social Change—
Man, the Master or the Slave

Revolutionary developments in technology and the accompanying changes in the society that have occurred since World War II are making drastic changes in the life of all of us. The existing patterns established in response to the situations of another day will impose very great burdens on the human beings living today. In our present social structure, human beings suffer extreme stress as technological change is imposed upon them. This whole process should be turned around. The question that really should be faced is how can man live a better life as a result of the new technological capabilities and social organizations that are now potentially available for his use.

Because of the radical changes now occurring and on the horizon, the process of evaluating what man can do in this situation is a very hard one indeed. The danger is that we will not grasp fully enough the opportunities that are afforded and will accordingly live too cautiously, keep too strong ties with the outdated patterns of the past, and have too little appreciation of the functions that are now available to us for the future. There are several areas in which technology offers tremendous opportunities or, if improperly viewed, tremendous threats, to the life of human beings during these periods of change.

First, the automation revolution. Automation promises to remove a lot of the menial production-line work that has been done by human hands over the last 50 years. It will mean a significantly shorter work week. It will mean that the demand for human participation in many production processes will be drastically reduced. It will mean that goods will be much more available to all of us. It will also mean that human beings will be very much detached from much of the production process. Our old ideas of a man's worth being measured in terms of what he can produce will have to be drastically changed.

Because of the continuing changes in technology, a workman will not be able to live his whole career doing one particular kind of job. An image held by each man of his ability to move to a new situation, to take on new responsibilities, and to make new contributions, will be vital even to the relatively unskilled laborer. New patterns of maintenance and distribution and repair will need to be developed in response to new situations.

Fortunately, the things that can be most readily automated are the very uninteresting, repetitive jobs in which men are trapped in the present industrial world. Our problem in this case is to develop a new set of capabilities in the people that now hold these jobs. If they can make the transition to the new kinds of life, their own lives will be much fuller. The transition, however, is not an easy one.

Second, the revolution in location of industry and management. Many of our large cities were built in response to the need for a large work force close to production lines and shops in the big industrial cities of our country. As new plants are built, automation is moved in, and as transportation becomes cheaper and more uniformly available to the population, there is a great shift away from the city. There is now no functional reason to have these industries clumped in one place. Rather, communities can be designed on the basis of providing the production capability while allowing the worker to have the most pleasant, enjoyable, and rewarding life. Similarly the management centers for industry were oftentimes in one or two big cities, like New York or Chicago. Now the pattern is to have a headquarters for the large companies scattered over the country. The exact location is less important than is the proximity to a large air terminal.

Third, the revolution in technical training. In the past

a one-job specialty for a professional man's lifetime was standard. A person who was interested in power engineering, for example, would expect to spend most of his productive life in that particular industry. Similarly, steel making or chemical process control or any number of such occupations, were life-long specialties. The tendency was to train people in the details of an area with the expectation that some 50 per cent of an engineer's time would be spent on really quite detailed problems peculiar to a particular industry. The situation now is quite different in that the time constant of a man's particular professional career may be no more than five or ten years, and not 40 or 50 years as it was in the past. Accordingly, the kind of training that is needed now does not go into the immediate details of a given situation, but rather must provide the man flexibility, so that he has the mobility to move from one job to another as the situation demands such moves. More important than the detailed training that a man receives must be the confidence that he has in his own ability to make changes when the need occurs.

Fourth, the revolution in communication and education. Twenty-five years ago the main method of communication was by reading from the printed page. A person was either closely in touch with the society or he was fairly isolated from it in its broad aspects, depending upon whether or not he could read well. The advent of television has very greatly changed the impediment to communication that is tied up with his ability to read.

The nature of the new communication capability makes us reevaluate the role of the teacher. In the past, the teacher mainly was teaching students how to read and write. She would bring a class of 25 to 50 students along a certain pathway, all together. They had to stay together, not because this was the best way for each to go through material, but because this was the only way that the teacher could keep in touch with all of them at once. Now, because of the ease of back-and-forth communication using computers, much greater flexibility can be afforded in the whole educational process. It is foreseeable that students will not even need to go to a central school facility. It is entirely feasible that small neighborhood units of the school system could be set up and could be kept in contact with the other units by means of audio and television links. Whether or not this is a good idea depends, not only on how usefully the teaching function can be realized with this arrangement, but also on the importance of other functions that the educational system should serve. For example, much social development comes to the students as a result of their education in the public school system. Probably the school serves a special social role. How important is it for community integration, for example? How important is it for a sense of belonging to the individual students? How important is it as a vehicle for enlarging the personal contacts of young people?

There is an even deeper question with regard to education. In the past there was a great deal of emphasis on training. The ability to read and write and to master arithmetic has always been considered of paramount importance. It will probably continue to have this importance. However, the question now comes as to whether or not one could put more time into development of the human potential in each of the students. It is quite possible that better training methods will allow the same kind of skill development, knowledge, and understanding to be transmitted to the students in a shorter time and accordingly that there will be time to realize some of the other functions which have had to receive a minor place in the past.

Fifth, technological advances have emphasized international ties. All nations, developed and developing, will gain substantially from cooperation rather than from isolation. There is a strong technological basis for the great increase in international ties that has taken place over the last twenty-five years. Many heretofore separate elements of the world have been closely tied together. Transportation, communication, and the feasibility of using high-speed computers have made it possible for the whole world to be knit together in a way that was not possible before. Now an operation can be analyzed carefully ahead of time. Significant alternatives can be checked out in detail and evaluated carefully. Planners and leaders can keep in touch with the whole operation.

Similarly we must look at the problems of world economic development from the vantage of the tremendous technological advancements. The problems of the world economic development are the problems of bringing technological advances and the possibilities for economic growth to all parts of the world.

While the United States is not interested in dominating the plans and lives of peoples of other nations of the world, we are very much interested in cooperative exchange between these countries. They can profit and we can profit by taking advantage of the complementarity of our own capabilities and theirs. Because we have much of the technology already developed, there are certain things we can do better than they. Because they have large untapped resources and manpower and material resources that are yet undeveloped, there are many things that they can do much more effectively than we.

Certainly, in every case, any development will have to be controlled and guided from the homeland in which it occurs. If we look back at our own period of development, we find that any of the attempts of European nations to guide or control us would have been met with very severe rebuff. By the same token, the development of the emerging countries will need to be largely generated at home. The dilemma posed by the world economic development problem is that intensive interactions are required from four different kinds of independent entities in the process which will bring revolutionary changes in economic, social, industrial, and governmental patterns for several generations. Centralized control of the development process is out of the question; and yet these four kinds of entities must be brought together to develop a more effective joint action.

The four entities are: (1) mature industries, (2) developing industries, (3) government entities in the developing countries, and (4) government entities in the developed countries. A common meeting ground will be indispensable as a prerequisite for interaction. Universities, in their primary role as observers, recorders, structurers, and extrapolators of culture and knowledge, will be in an ideal position to provide this common meeting ground.

The Environment of Continuing Technological Change

A transition from the past to the present is not a one-step transition. There are continuing cascades of technological advances that will be offered to our society into the indefinite future. The response of the society, therefore, is not to make one transition and then settle again into a new pattern of more or less static behavior. A continuing look into the future will be required for our society. Functions which must be performed to aid society in adapting to the future possibilities will need to be in operation continuously and over a long period of time, many generations, in fact. Because of the long-term prospect for change, a set of permanent agencies probably should be developed in order to realize the advantages that the changes will present.

The Integrative Nature of New
Technological Opportunities

Because of the advances in transportation, communication and data processing skills, the outreach of each individual activity is much greater than it was before. If we view each individual activity as a point on a plane, then the scope of influences may be thought of as circles surrounding those points. As the range of interactions increases, the diameters of the circles can be increased to represent this increased scope. Now we see a great overlapping. This is a source of great possibility or great confusion. As an example of this increase in scope of activities, one can mention the use of an automated delivery system. A given company, for example, could cut down its inventory cost by having a high-speed automated delivery system and only one warehouse. The Air Force has, in fact, instituted such a system for the delivery of jet engines by air to any place in the world whenever they are needed; thereby achieving a very substantial overall operational savings.

When a private industry tries to do this same thing, many problems immediately arise because so many independent entities have to be coordinated. There has to be a different kind of a packaging system, a high-speed ground delivery system, and a greatly modified air traffic control system. These entangling interactions could be handled if they were properly coordinated, and in many cases we find that such coordination is economically profitable. In other cases, however, the interactions may be too high, and technological capabilities might not be operationally feasible to exploit.

The need for integrative system planning occurs in many different kinds of situations. It obviously comes up in the educational system, particularly in the connection between education and communications. It comes up in medical care planning where there are tremendous advantages to be gained by having medical records centrally kept and many hospital operations monitored by electronic data-processing equipment. The automation of the banking system is causing a revolution in the ability with which individuals can obtain credit over wide geographical regions. Advertising on one hand and shopping by individuals on the other will be very greatly simplified by the communication capability that will be available.

How to Exploit Change

Revolutionary technological changes and consequent changes in our society are already in motion. Man can either be the master of the benefits that these changes promise, or he may be the slave. If man is to be the master, more than a superficial understanding of the changes and their consequences must be clearly grasped.

The traditional free enterprise system has served our country very well in its developments in the past. Freedom begets initiative, variety, and competition, all of which are extremely valuable. Unfortunately, the present free enterprise system alone does not provide the communication of information nor does it provide adequate means for initial exploration that is needed to harness new technological advances. What is needed then is to add new functions to the present capabilities of society so that we can gain both the fruits of free enterprise and the benefits of new technology.

There are three separate functions which must be provided our free society so that it can adapt to change and exploit it:

1. Broad projection into the future must be widely available.

2. Broad education and training must be generally available to all people.

3. An integrative sanctuary to allow preliminary exploration of future developments must be developed.

The future is a combination of the treasures of the past and fruits of the new. Because many new technological advances have sweeping consequences and require extensive development, society must be given the scope of understanding and the lead time necessary to exploit them. Exploration is obviously necessary. It is not efficient nor even feasible for each enterpriser to do the exploration by himself. A common exploration of the future, open and widely available, is indispensable.

Broad education and training for the new day must involve two separate functions. An initial education and training base must be provided the youth so that they will have the breadth and the flexibility to cope with the variety and uncertainty of the future. They must have a self-image of mastery of the future so that new opportunities are exciting rather than threatening. Second, there must be broad orientational training so that planners, designers, managers, and operators can work effectively in the new positions which will be continually arising. This orientational training is needed not only by the youth but also by the mature person who must have the mobility to make the transition from an obsolete job to a future-oriented opportunity.

Technology offers new opportunities that are both radical departures from the past and are unusually broad in scope. At the outset of the development of an area that requires the participation of many independent entities, such as a nationwide system of air freight or a large educational program, open and fairly detailed exploration by many independent entities should take place in a sanctuary away from the marketplace so that opportunities can be assessed, alternatives can be explored, benefits can be estimated, and rough developmental strategies can be developed. After the initial exploration, the efficiency of the marketplace and the benefits of competition can be restored. The initial policy analysis of a new area properly belongs in the sanctuary. Developmental and operational decisions are usually better made in the marketplace.

We pass now from the new functions to be performed to a consideration of the nature of the agency or agencies which might perform them. We then look fundamentally at intellectual activity (not at universities) and then infer what the role of the university might be in the future if it is to have a role in this area.

Two Indispensable Characteristics

There are two characteristics of any agency or institution which will be indispensable for the success of that agency in developing the opportunities that technology provides our society. First, the agency must combine the two basic drives of our society—the basic drive *to do* which has characterized our industrialists, our managers, our enterprisers, and our politicians and has made our country as strong as it is today; and the drive *to understand* which has characterized the efforts of scholars through the ages. The present situation calls for a *combination* of the drive *to do* with the drive *to understand.* The revolution that we are involved in is so drastic that doing without understanding will be futile. The opportunities are so great that understanding without doing is equally pointless.

Second, the agency must have the capability for regeneration. Though the present revolutionary forces for change are strong, our culture has tremendous natural inertia and must provide sustained guidance and direction for several generations if it is to foster long-term planning and development. If one looks back at the institutions that exist today, those with the longest history and the greatest sustained influence are in the universities. It therefore follows that if an agency is to be effective for long periods, it must involve the capability for regeneration which is uniquely found in the universities.

In contemplating an agency or activity to help develop the potential that the new technology provides our society, we see that any workable agency which we can plan must couple the drive to do with the drive to understand in order to be effective. It must also possess the capability for regeneration in order to have a long enough lifetime to be effective.

A Model of Intellectual Activity

At this point, it is well to look at a model of intellectual activity. It can be characterized in terms of layers, like layers on an onion. In the central core of intellectual activity is the work in the classics, followed successively by layers in the work of humanities and sciences, the work in the professional disciplines, the work in planning institutes and in research and development laboratories, and an outer layer of the operational world, which involves governmental and private institutions. Each of these layers is characterized in terms of a time constant. The time constant is the equivalent of a half-life of an activity. The time constant of the classic core is greater than 100 years; the time constant of the humanities and science program is like 50 years; the time constant of the professional schools can be derived on the basis of the function that they serve: a professional school must train its students in such a way that their training will be meaningful to them when they are at the peak of their careers, which is about 20 years after they have left the university. Accordingly, the programs that we give them have to have time constants of the order of 20 years. Planning institutes and research and development laboratories have time constants on the order of five years. In the operational world, the time constants are on the order of two years.

It is very important to notice that each layer is influenced very strongly by its adjoining layers. Generally, the drive to do (from the real world) is fed in from the outer layers toward the center, and philosophical structuring is supplied from the inner layers to the outer layers. Thus, the basic drive to understand comes from within and the basic drive to do comes from without. In the time of very high technological stress, which is a characteristic of the present time, there is a great tendency for these adjoining layers to come detached from each other. It is very painful for the academician to try to structure the new and ever-changing situation in the real world. His whole observing, measuring, and evaluating mechanism is tremendously upset by the tumultuous changes that are occurring in the outside world. The opportunities to the operator on the outside are so great that he may be tempted to take on projects which in the short-term look very meaningful but which may in the long-term not serve any very useful function. Thus, he would ignore the inner layers.

In an environment of great technological opportunity, as we have now, it is important that these layers be closely coupled. We notice, for example, that the humanities and science programs in the university can be greatly aided in getting signals from the outside world by a proper collaboration with the professional schools. It serves their mutual best interests to work together on this program rather than independently. Any good operation of a given layer would involve at least the two adjoining layers, and possibly the two outside of those.

The University Must Be Involved

Several functions must be served in order to enable man to exploit properly the changes in technology and society which are coming ever faster. Two of these have traditionally been the function of the university. One could be also assumed by the university. As the collector, organizer, interpreter, and disseminator of our culture, the university is in an ideal position to forecast the future. The extent to which the future can be predicted depends on the lasting things that exist in the present, and the university, in its role, is almost uniquely able to serve this function.

In providing the initial training for the young and in developing the basis for mobility that is needed for those in mid-career, so that they can adapt to new technological changes, the university would be serving its traditional role in education, but in a somewhat new way.

In providing an integrative sanctuary, the university would be taking on a new function. In order for it to provide a sanctuary, it will be required to bring the best sources of understanding and data on new technology right to the university campus where the incorporation of these ideas into the university program will be direct and effective. The objectivity and detachment that has typically characterized university people when they view the world of affairs, is absolutely essential for the sanctuary environment. Not only *can* the university serve these special functions well, but it *must* serve them if it is to also effectively accomplish its traditional role in the future.

As was indicated in the discussion of the model of intellectual activity, in an environment of high, drastic technological change, it is very easy for the adjoining layers in the intellectual structure to become detached from one another. If the university becomes involved to the extent of taking term responsibility in helping to project the future, in serving as a sanctuary, and in training people to meet the new challenges and opportunities that technological change bring about, it will be sufficiently involved in the world of affairs that the signals that come to it will be strong and meaningful.

The university *can be* effectively involved and the university *must be* involved in the world of affairs in order that man can exploit the technological advances that are coming into our life today.

If the university were to take on the new role and abdicate the old, society would be the loser both because the old role is still very much needed and the new role can't be served by the university alone. In this new role the university must be coupled to the world of affairs in an appropriate way. With such a coupling it can contribute simultaneously to both roles.

An Institute for Technology and Society

I propose that an Institute for Technology and Society be developed at the interface of the university and the world of affairs so that the university can properly serve the functions that society needs if it is to be capable of exploiting future technology. In order to determine the nature of this institute, we must take a more detailed look at the projective planning function.

Planning Must Be Interdisciplinary

Perhaps this very short description of the layers of intellectual activity and its logical consequences can be taken to be a prescription for planning work that is to be interdisciplinary in the broadest sense. In what sense is interdisciplinary work broadly interdisciplinary? First, in the academic sense. Many academic disciplines must be simultaneously involved. Physical sciences and behavioral sciences must be jointly involved, for example, in planning transportation systems, or in planning communications systems, or in planning educational systems. Very often it's important to understand the political and human issues involved before one undertakes technological innovations.

Second, in the professional discipline sense. The artificial barrier that has often been raised between management and engineering must be eliminated. Generally, the manager knows what is technologically available, but he doesn't know what is technologically feasible. Analogously, the engineer very often knows what is technologically feasible, but he doesn't know what is operationally feasible. There need to be involvements between engineers and educators, between educators and lawyers, between managers and each of the other three. For example, the technological impact on the medical profession has brought a great need to bring management and engineering and medicine much closer together than they have been in the past.

Third, in collaboration among what have been heretofore private institutions. For example, it will be necessary for computer companies, transportation companies, packaging companies, and many manufacturers who would use high-speed ground transportation to cooperate much more closely if the high-speed transportation system problem is to be effectively attacked. In planning educational systems, it is no longer possible for textbook publishers, manufacturers of audio-visual equipment, television manufacturers, and computer manufacturers to operate separately.

Fourth, in collaboration among different public institutions on the same project. One of the clearest examples of this in recent times is the need for people in the State Department and in the Department of Defense to be working jointly on the problems of limited war. Further, we see that the distinctions between military activity and economic activity can be very artificial these days. Similarly the involvement of budget planners, educators, and communication engineers, for example, is clearly needed in the development of educational television.

The few foregoing examples have indicated the broad scope of interdisciplinary activity. The problems of establishing interdisciplinary activity are very great indeed. There must be some set of special mechanisms developed to induce different agencies and different groups to work together effectively.

Mechanisms of Achieving an Interdisciplinary Program

There are a number of factors that are important in easing the difficulties in establishing interdisciplinary work.

1. The program must be focused on specific problems. The fact that a specific set of issues can be focused upon makes it possible for individuals with different backgrounds to work together toward an enriched understanding of the problem. The differences in backgrounds would be a great impediment to mutual understanding without the focusing that is provided by the specific issues.

2. Alternation must take place between the pressure to do and the pressure to philosophize. It is important to distinguish these pressures. Any group or team has to be motivated, at some point, by the pressure to do. Specific decisions must be made, specific plans must be carried out, specific delegations of authority have to be made, resources have to be committed. Alternating with this pressure to do is the pressure to understand. Each time a long-range plan is studied, it must be reviewed in breadth and detail. There must be an understanding of the deeper issues involved if the program is to be successful.

3. A sanctuary must be provided away from the pres-

sures of the world. A combination between field agencies and university agencies would allow professors and students to work for extended times in the field, becoming involved in the pressure to do, and then returning to the university with the philosophical issues more clearly drawn. Similarly, senior fellows can come from the practical world to the university to structure new problems.

4. A neutral meeting ground must be provided for competing or contending real-world elements. A sanctuary must be provided where integrative planning can illuminate issues before the competition necessary for efficient operation raises barriers between these elements. The detachment of the academic institute would be a great aid to communication among the contending elements in an operational system.

A Plan to Couple the University to the World of Affairs

Our premise is that the university has a vital role in helping man exploit the changes in technology. To make the university role meaningful and productive in the fullest sense, the university parts must be coupled to the world of affairs so that continuing interchanges can take place.

We view a set of layers of activity much like the layers presented in Figure 1. The activities toward the center are closer to the intellectual core. The layers toward the edge are closer to the world of affairs.

The integrative possibilities for the institute for technology and society are pictured in Figure 2. Its most important function would be to provide a common meeting ground for academic disciplines, for professional disciplines, for public institutions, and for private institutions.

The plan of operation is schematically represented in Figure 3. The left-most column represents the foundation

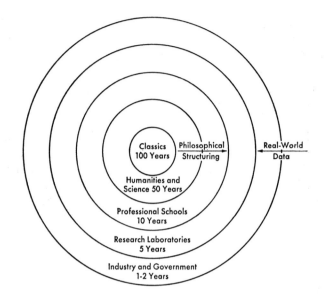

FIGURE 1

academic disciplines with which the professional schools have strong connections. Although the most important of these is mathematics; economics, physics, political science, psychology, and anthropology are all areas that are relevant.

The second column represents professional disciplines, with the upper part representing established disciplines and the lower part representing the new activity that has been added in systems. This will be described in more detail later.

The third column represents the coupling activity to the world of affairs. The institute will be involved in working on specific problems of society and will form a strong coupling link between the academic activities, represented by the two left-hand columns, and the field agencies which are represented by the right-hand column. The right-hand column is divided into three parts, government agencies, nonprofit institutions, and industrial companies.

The structure of the program in Engineering-Economic

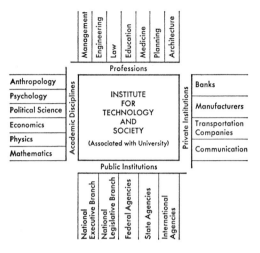

FIGURE 2

Systems is presented in Figure 4. The upper half of the figure represents the "theory" and the lower half represents "the application." Interaction between theory and application is even more critical for systems than it is for the more established professional disciplines. The systems concept and

Academic Disciplines	Professional or Applied Disciplines	(To Couple)	Practical Entities
Mathematics	Established Professions	Institute for Technology and Society	Government Agencies
Economics	Engineering		
Physics	Law		Non-Profit Institutions
	Medicine		
Political Science	Business		
	Education		
Psychology	Engineering- Economic Systems		Industrial Companies
Anthropology			

FIGURE 3

methodology are meaningful only to the extent that they apply generally to a wide variety of substantive areas. Continual application to new areas with continuing evaluation, interpretation, and modification of concepts is vital.

SYSTEM CONCEPTS AND METHODOLOGY

Mathematical Concepts	Problem-Solving Concepts (System Analysis)
Modeling	Decision Analysis
Optimization	Policy Analysis

APPLICATION OF SYSTEM ANALYSIS TO SUBSTANTIVE AREAS

TECHNOLOGY		SOCIETY
Education		Urban Development
Medicine	Integrative Study of Man	Regional Development
Transportation		Public Economics
Communication		
Water Resources		
Power Systems		
Banking		

FIGURE 4

Mathematics is the language of system analysis. The two primary mathematical activities for systems analysis are modeling and optimization. Modeling is associated with arriving at a description and an analysis of systems as they are as well as what they are projected to be. Optimization is an essential component of design, in that the best or most attractive alternative must be chosen. Although mathematics as a language is sharp, clean, and precise by itself, it often lacks the color, shading, and tone to represent situations in the real world. The approach to problem solving must be added to mathematical language to make system theory complete.

Two important aspects of problem solving involve the setting of policy and the making of decisions. Policy issues involve the determination of the scope and objectives in a given situation. A decision is the irreversible commitment of resources. Under the usual practical conditions of uncertainty and dynamic interaction, risk and time preferences must be accounted for. Usually in strategic situations, that is, in macrosystem problems having a large scope in both time and space, policy issues dominate. In tactical situations, that is, in microsystem problems involving detailed constraints and multiple interactions, decision questions dominate. Every practical situation involves some policy issues and some decisions.

In all practical areas, system activities must serve man; human needs, human goals, human potential must be realized. Technology and the various institutions of society are man-made instruments whose only justification is to serve man. The outreach of the systems profession is through the applications areas. Work in humanities and sciences provides a foundation for technology and the study of social institutions and, hopefully, a deep appreciation of the basic needs of man. In addition to the basic physical needs for food, shelter, and clothing there are the limitless possibilities for human development. Not only is man as an individual important, but mankind as a group must be considered.

When actions are to be taken to achieve specific goals, the connection to the world of affairs is obvious. Actions in the various technical areas are integrated through social, political, and industrial organizations into a coordinated unit.

A Pilot Program at Stanford University

When one attempts to understand the scope of the issues that are involved in long-range planning today, he runs a risk of being misunderstood, as being too ambitious, or

too arrogant, or too self-assured. It would be very unrealistic
for us at Stanford University, in the small program that we
have, to claim a really meaningful long-term plan or an institution which has a good chance of carrying it out. Our
hope is that we can understand what the problems are and
that we can do a few pilot studies to verify or reject our assumptions about the state of the world. We have a few
ideas as to how we would carry out a combined academic
and operational program. Although our operation is small,
it has been going on for about five years. We have completed
a number of combined practical and analytical projects. We
have developed a number of academic courses. Though the
progress we have made is small compared to that needed to
do the job that we are contemplating, nonetheless, we believe we have some useful data.

We have had about 40 interns working on problems in
the field; we now have about 80 doctoral students involved in
our program, of whom about 50 are enrolled in the Engineering-Economic Systems Department. Most of our work over
the last five years has been devoted to the development of a
strong analytical program in the academic aspects of our
work.

With regard to applications of system analysis to the
substantive areas, we have a great need to form much stronger
connections with the humanities and sciences. If engineering-economic systems is to be really meaningful as a profession, it must be built to serve the needs of man. At present, society has far too little understanding of what man's
needs are; we hope to join professors from the humanities
and the sciences to explore and define the needs of man in
practical situations.

We have a number of practical projects under way in
the field. First, we are carrying on a long-range educational
planning project for the Bureau of Research in the Office of
Education. Second, we are doing a problem in conjunction

with the Stanford Hospital on long-term medical care planning and hospital operation. Third, we have had a fairly extensive set of internships in Washington, including work with the Department of Commerce and with the Department of Transportation on long-term transportation planning. This is perhaps the strongest single element of our program. It was developed with the help of Ford Foundation funds. Fourth, we have a few individual research projects under way on regional economic development, particularly transportation studies in Brazil and Chile; in connection with the Bonneville Power Administration; with the Institute of Electrical Development in Mexico; and with General Electric, Westinghouse, and Jones & Laughlin Steel Company. We have done several studies of electric power system planning and operation. We have developed a strong relationship with several of the consulting companies, particularly with the Stanford Research Institute. Finally, we have established a small program in Water Resource Development.

Overall Plan for Operation of the Program

The effort of professors and graduate students in the Engineering-Economic Systems program is divided into three parts. Something more than a third and less than half of the activity in Engineering-Economic Systems is tied up with the developing of core concepts for system analysis and doing research on basic concepts and methodology. The comparative studies on the usefulness of the various analytical concepts and methodologies for examining problems in various practical areas is an indispensable part of the program. It is important to try our concepts of operation on both tactical and strategic problems. We find that a given program begins with a long-term projection. Then comes policy analysis, followed by the making of plans and the developing of programs where allocation of effort is actually made. Finally, funds are committed, the development process takes place, the design study is carried out, and the installation and operation is accomplished.

Generally, we think of macrosystem problems as being those which are involved with the long-term planning. In this case, the problems are those of settling on objectives and getting the structure of the overall system clear. Microsystem problems are concerned with design and operation. Generally, as one gets toward the operational problems, the models become definitive, the constraints become more important, operational data are more readily available, the models can be verified on the basis of practical experimentation, and so on.

If the language and methodology and concepts are meaningful, they should be applicable to a wide spectrum of problems, from those that occur in private industry to those that occur in long-term governmental planning. These provide a very rich field in which comparisons can be made.

With regard to field work, a project is usually started because some professor or some set of people from the outside are interested in going into a given area. One of the functions of our institute for technology and society is to provide support for professors who wish to explore the possibilities of a new area. They talk with practitioners who are working in relevant areas and consult with government officials who have responsibilities or potential responsibilities in the area that is represented by this work.

After the initial exploration has been made, study seminars are oftentimes set up within the institute for graduate students and visiting fellows. Very often experts are brought in on a visiting basis for a short time to work with us at the university.

Finally, as the area becomes more clearly developed, an exploratory project is usually set up for six months to a year during which time an initial study of the problems is made. These exploratory projects are set up with visiting fellows from government and industry and with professors from the relevant disciplines, either from Stanford or from other universities, and with a few of our advanced graduate students. After exploratory studies and projects have been done, field work is often undertaken with students serving as interns in the area in which the field work is relevant. Since our interns are well-trained graduate students who work under only very limited supervision and are responsible on a term basis in the field, they represent a strong coupling between our work at the institute within the university and with the work in the field.

Our aim is to keep the work of the institute strictly on an exploratory basis. As soon as a project is well enough formulated, it can be taken over by some nonprofit corporation or some government or private laboratory. We are very anxious to have this happen. Our function is best served by keeping the university operation quite small and very closely tied up with the academic structuring operation. Very often our professors and students take term responsibility in the field where they serve as consultants to field agencies or industries or government, or as interns, if they are students. We are very serious about keeping the average age of those long-term participants in our operation under 30. If we do this, then we have to have a very small permanent staff and have to rely mainly on visitors and professors who take term responsibility within the institute program.

Summary

The function of this paper has been to lay out the scope of activity that must be developed so that the advancing new technology can more fully serve the needs of man. The needs are great, the challenges are exciting. Many exploratory projects such as ours will be necessary to develop a workable long-term program. I am sure many of us in industry, government, and the universities are eager to become a part of it.

Notes on Contributors

DANIEL ALPERT

Dr. Daniel Alpert is dean of the Graduate College, University of Illinois, Urbana, Illinois. He was professor of physics and director, Coordinated Science Laboratory, University of Illinois, Urbana, from 1957 to 1965. Dr. Alpert was associated with the Westinghouse Research Laboratories, East Pittsburgh, Pennsylvania, from 1941–1957; as a research fellow, 1941; research physicist, 1942; manager of the physics department, 1950; and associate director of the laboratories, 1955. Dr. Alpert was born in Hartford, Connecticut on April 10, 1917. He was awarded the B.S. degree from Trinity College in 1937 and the Ph.D. degree from Stanford University in 1941.

JACOB E. GOLDMAN

Dr. Jacob Goldman is director, Scientific Laboratory, Ford Motor Company. He joined the Ford Motor Company in 1955 as manager of the physics department, became associate director of research in 1960, and assumed his present position in 1962. He was a research physicist in the research laboratories of the Westinghouse Electric Corp. from 1943 to 1950 and a member of the faculty of Carnegie Institute of Technology from 1950 to 1955. Dr. Goldman was born in Brooklyn, New York, on July 18, 1921. He received his A.B. degree from Yeshiva College in 1940 and his M.A. and Ph.D. in physics from University of Pennsylvania in 1943.

WILLIAM K. LINVILL

Dr. William K. Linvill is chairman of the Department of Engineering-Economic Systems at Stanford University, Palo Alto, California. He joined Stanford University in 1960 as a professor of electrical engineering, became chairman of the Institute in Engineering-Economic Systems in 1963 and assumed his present duties in 1967. Dr. Linvill was a member of the faculty of the Massachusetts Institute of Technology from 1949–1958, as an assistant professor, 1949, and associate professor, 1953. He served for two years, 1956–1958, while on leave from M.I.T., with the Institute of Defense Analysis. He became a senior staff member of the RAND Corporation in 1958. Dr. Linvill was born in Kansas City, Missouri, on August 8, 1919. He was awarded a B.A. degree by William Jewell College in 1941, and a joint B.S.-M.S. degree and a Ph.D. from Massachusetts Institute of Technology in 1945 and 1949, respectively.

GEORGE E. PAKE

Dr. George E. Pake is executive vice-chancellor and provost and professor of physics at Washington University, St. Louis, Missouri, a post he assumed in 1962. Dr. Pake was born in Jeffersonville, Ohio, on April 1, 1924. He was awarded B.S. and M.S. degrees by Carnegie Institute of Technology in 1945 and the Ph.D. degree by Harvard in 1948. He was associated with Washington University from 1948 to 1956; as assistant professor, 1948; associate professor, 1952; professor, 1953; and chairman of the Department of Physics, 1952. He was professor of physics, Stanford University, 1956 to 1962. He was awarded honorary D.Sc. degrees from University of Missouri, Rolla, and Carnegie Institute of Technology in 1966. He is presently a member of the President's Science Advisory Committee.

EMANUEL R. PIORE

Dr. Emanuel R. Piore is vice-president and chief scientist and a member of the board of directors of International Business Machines Corporation. He joined IBM in 1956 as director of research and was elected a vice-president in 1960. Dr. Piore was associated with the Office of Naval Research from 1946 to 1955, serving as chief scientist for the last four years of this period. Prior to joining IBM, he was vice-president for research of the Avco Corporation. He is a member of the National Science Board and a former member of the President's Science Advisory Committee. He is a trustee and member of the Executive Committee of the Sloan-Kettering Institute for Cancer Research, chairman of the Committee on Scientific Policy of the Memorial Sloan-Kettering Cancer Center, and a trustee of the Woods Hole Oceanographic Institution. Dr. Piore was born in Wilno, Russia, on July 19, 1908. He received his A.B. and Ph.D. degrees in physics from the University of Wisconsin in 1930 and 1935. He served as an instructor at the University of Wisconsin from 1930 until 1935. He received the honorary degree of D.Sc. from Union University in 1962, and from the University of Wisconsin in 1966.

WILLIAM J. PRICE

Dr. William J. Price is executive director of the Air Force Office of Scientific Research (AFOSR), currently on sabbatical leave as a federal executive fellow at The Brookings Institution, Washington, D. C. Before coming to AFOSR in 1963, Dr. Price spent six years with the Aerospace Research Laboratories, OAR as chief scientist. Dr. Price was born in Alexandria, Ohio, on December 3, 1918. He received his A.B. in 1940 from Denison University, and his M.S. and Ph.D. from Rensselaer Polytechnic Institute in 1941 and 1948 respectively. He did research at RPI where he was a member of the physics department and also at Bendix Corporation, Battelle Memorial Institute, and the Air Force Institute of Technology where he was head of the Department of Physics.

CHALMERS W. SHERWIN

Dr. Chalmers W. Sherwin is an independent consultant in the fields of research and development managment and in scientific and technical information. He is currently leading a special task group on the compatibility of national systems reporting to the director, Office of Science and Technology, Executive Office of the President. He was Deputy Assistant Secretary of Commerce for Science and Technology in 1966. From 1963–1966 he was Deputy Director of Defense Research and Engineering responsible for the areas of research, electronics, chemical technology, materials, and laboratory management. From 1960–1963 he was vice-president and general manager of the Laboratories Division of the Aerospace Corporation in Los Angeles. Dr. Sherwin was born at Two Harbors, Minnesota, on November 27, 1916. He was graduated in 1937 from Wheaton College, Illinois, with a B.S. degree and received his Ph.D. in physics from the University of Chicago in 1940. He was an assistant in physics at the University of Chicago in 1941 and a member of the staff of the Radiation Laboratory at M.I.T. from 1941 to 1945. After one year at Columbia University, he joined the physics department of the University of Illinois as an assistant professor, becoming associate professor in 1948, professor in 1951, and associate director, Coordinated Science Laboratory, in 1959. During 1954–1955, while on leave of absence from the University of Illinois, he served as chief scientist of the U.S. Air Force.

MORRIS TANENBAUM

Dr. Morris Tanenbaum is director of research and development (assistant administrative officer) at Western Electric Company's Engineering Research Center at Princeton, N.J. Dr. Tanenbaum began his Bell System career in 1952 as a member of the technical staff of the Bell Telephone Laboratories in Murray Hill, N.J. He was advanced to head of the physical chemistry of solids department in 1956 and was named head of the research on the crystalline state department two years later. He was promoted to assistant director of the Metallurgical Research Laboratories in 1960 and to director of the Solid State Device Laboratory in 1962. On August 1, 1964, Dr. Tanenbaum accepted his present position with Western Electric Company. He was born on November 10, 1928, in Huntington, W.Va. He received a B.A. degree in chemistry from Johns Hopkins University in 1949 and the Ph.D. in physical chemistry from Princeton University in 1952.